A Study on Laser Forming Characteristics of
High Strength Steel by Finite Element Analysis

基于有限元法的高强度钢激光弯曲成形特性研究

张雅晶　董文彬　著

西北工业大学出版社

西安

图书在版编目(CIP)数据

基于有限元法的高强度钢激光弯曲成形特性研究/张雅晶,董文彬著.—西安:西北工业大学出版社,2019.12
　ISBN 978-7-5612-6781-3

　Ⅰ.①基… Ⅱ.①张… ②董… Ⅲ.①激光技术-应用-高强度钢-弯曲成型-研究 Ⅳ.①TG142.7

中国版本图书馆 CIP 数据核字(2019)第 293595 号

JIYU YOUXIANYUANFA DE GAOQIANGDUGANG JIGUANG WANQU CHENGXING TEXING YANJIU
基于有限元法的高强度钢激光弯曲成形特性研究

责任编辑:王　静	策划编辑:付高明
责任校对:孙　倩	装帧设计:李　飞

出版发行:西北工业大学出版社
通信地址:西安市友谊西路 127 号　　邮编:710072
电　　话:(029)88491757,88493844
网　　址:www.nwpup.com
印 刷 者:西安日报社印务中心
开　　本:787 mm×1 092 mm　　1/16
印　　张:4.875
字　　数:128 千字
版　　次:2019 年 12 月第 1 版　　2019 年 12 月第 1 次印刷
定　　价:20.00 元

如有印装问题请与出版社联系调换

Preface

Laser forming is a novel material processing technique which requires no expensive dies and no mechanical contact. It is a complex transient thermal-mechanical coupling process. The deformation of the material is induced by the residual plastic strain which is caused by thermal stress depending on the laser scanning. Thus, laser forming is available for almost all kinds of hard forming materials, such as high strength steel, titanium, and composite materials and even ceramic.

Namba was the pioneer who applied laser forming process on bending metal sheet in 1984. Since then, significant interests have been attracted by the researchers all over the world. Earlier researchers like Vollertsen F. focused on the mechanisms of laser forming. It is generally suggested that there are three kinds of mechanisms in laser bending process, defined as temperature gradient mechanism (TGM), buckling mechanism (BM) and upsetting mechanism (UM), respectively. Since 2000's, more and more researchers have studied the effects of laser parameters, material properties and specimen dimensions on temperature and bending deformation, such as Cheng P. J., Shi Y. J., Shen H. and so on. Some of them have also studied edge effect of laser forming and achieved fruitful results. The research methodology on laser forming process falls into three major categories: experimental method, analytical method and finite element method. The experimental method is more reliable but time-consuming and costly. The analytical method claims large and fussy calculation. Comparatively, the finite element method is an intuitive and convenient method which saves time and cost greatly.

In this study, the finite element method is employed to investigate the laser forming characteristics of dual phase 980 high strength steel. The research background and laser forming mechanisms are introduced in Chapter 1 and the theory of finite element method is described in Chapter 2. In Chapter 3, a 3D finite element model is built to simulate the laser forming process and an experiment under the same conditions is carried out to verify the reliability of the finite element model. In Chapter 4, the effects of laser parameters and specimen size on bending angle in the single scanning laser forming process are analyzed using the built finite element model. The scanning pattern effects in the two-time and three-time laser scanning are presented in Chapter

5. Chapter 6 gives the conclusions and recommendations of this study.

In this text, Chapter 1, 2, 5 and 6 are written by Yajing Zhang of Anhui Science and Technology University, while Chapter 3 and 4 are completed by Wenbin Dong of Anhui Science and Technology University.

Any comments, suggestions or corrections regarding this text would be gratefully accepted by the authors.

Author

Feb., 2020

Description

Laser forming is a newly developed technique that uses a defocused laser beam to bend metal sheet by thermal stress rather than external force. The bending shape and position are determined by the factors which include the laser power, scanning speed, and beam size. Most research work on laser forming to-date has been focused on the materials with good deformability experimentally and theoretically, such as mild steel, aluminum, titanium, and so on. A few studies on finite element method (FEM) simulations of laser forming process have been found but most of them used the square heat source or small stroke jumping heat source model. In this study, a series of FEM analyses using Gaussian-distributed heat sources are performed to investigate the laser forming characteristics of high strength steels which are widely used in industry but difficult to be deformed.

The first part of this text describes a series of single-time straight-line laser forming simulations for DP980 high strength steel using ABAQUS software. In the simulations, a nonlinear transient three-dimensional heat transfer finite element model is developed and the heat source from laser beam is defined by the user subroutine DFLUX. Through the comparison between finite element analyses and experimental results under the same conditions of laser forming process, the simulation results are verified and the temperature distributions and final bending angles are presented.

The second part of this text presents the finite element analyses of one-time laser scanning process. The laser parameter (i. e. laser power, laser scanning speed and line energy) effects and specimen sizes (i. e. sheet length, width and thickness) effects on forming characteristics of DP980 high strength steel are investigated.

In the third part of the text, finite element analyses of multi-time laser scanning processes are performed, which are divided into two-time scanning and three-time scanning. Through the analysis of temperature distribution and final bending angle, the effect of scanning pattern on forming characteristics is confirmed.

The conclusion is that the finite element analysis using the developed finite element model is verified by the experimental work. The finite element model contributes to a better understanding of the laser forming technique, and the effects of the impact factors in a laser forming process are able to be investigated efficiently.

Contents

1 Introduction ··· 1

   1.1 Research Background ·· 1

   1.2 Laser Forming Mechanism ··· 3

   1.3 Research Trends and Applications ··· 9

   1.4 Research Motivation and Objectives ·· 11

   1.5 Outline of the Paper ·· 12

2 Theory of Finite Element Analysis ·· 13

   2.1 Introduction ··· 13

   2.2 Finite Element Method (FEM) ··· 13

   2.3 Thermal Problem ··· 14

   2.4 Structural Problem ·· 17

   2.5 Summary ·· 21

3 Finite Element Modeling and Experimental Verification of Laser Forming Process
·· 22

   3.1 Introduction ··· 22

   3.2 3D Finite Element Modeling ·· 22

   3.3 Experimental Verification ·· 29

   3.4 Summary ·· 30

4 Finite Element Analysis of One-time Laser Scanning ···································· 32

   4.1 Introduction ··· 32

   4.2 Analysis of Laser Parameter Effect ·· 32

   4.3 Analysis of Specimen Size Effect ·· 37

   4.4 Summary ·· 41

5 Analysis of Scanning Pattern Effect in Multi-time Laser Scanning ·············· 43

   5.1 Introduction ··· 43

5.2	Two-time Scanning Analysis	43
5.3	Three-time Scanning Analysis	45
5.4	Summary	49

6 Conclusions and Recommendations ··· 50

6.1	Conclusions	50
6.2	Recommendations	50

Appendix A Overview of Laser ··· 52

A.1	Properties of Laser	52
A.2	Types of Laser	53

Appendix B Lists of ABAQUS Computer Program ··· 54

B.1	Finite Element Analysis of One-time Laser Scanning Process	54
B.2	User Subroutine for the Gaussian-distributed Heat Source Model	63

References ··· 65

Acknowledgement ··· 69

1 Introduction

1.1 Research Background

Forming techniques, in a broader sense, comprise a variety of metal working processes in which the material is shaped in solid state by plastic deformation. In the present context, forming process is primarily referred to as sheet material forming such as bending. Conventional mechanical bending process for a sheet material involves a set of bending die and punch with a sheet material placed between them. During bending, the sheet material is plastically deformed into desired shapes by application of suitable forces such that the shape of the sheet material conforms to the contours of the die and punch. However, due to the high cost of dies and punches, the conventional mechanical sheet material forming processes are primarily suitable for mass production and they are uneconomical for a rapid production of few parts such as those required for test prototypes and special shapes. Also, the conventional mechanical sheet metal forming processes are often associated with the inherent effects such as spring back effects where the actual bending angle is always less than the desired bending angle defined by the dies.

The idea of using lasers for forming of sheet material was first conceived in 1983 by Kitamura[1] in Japan who used a 15 kW CO_2 laser to bend 22 mm thick steel plates successfully. Since then, significant interests have been attracted toward use of laser for forming of sheet materials, especially sheet metal. Figure 1.1 illustrates the schematic diagram of a straight-line irradiation process which produces a bending angle from a flat sheet metal piece by a defocused laser beam[2]. The sheet metal is clamped at one side on a computer numerical control (CNC) machine. The heating on the material surface by a laser beam occurs on one side along a selected line while the CNC table is moving. The sheet metal expands in the heated zone and thermal stresses are produced by the restriction of the surrounding material. Then, the thermal stress will lead to a bending angle in the sheet metal. The surface melting of the sheet metal is avoided by adjusting the laser parameters such as laser power, scanning speed and beam diameter. Industrial lasers used as a beam source include excimer, carbon dioxide (CO_2) and neodymium: yttrium-aluminum garnet (Nd:YAG). The choice of the laser source depends on the task. The excimer laser with the wavelength band of 193 ~ 351 nm can be used for the production of parts with length and width dimensions in the range of 10 μm due to its extremely short pulse duration. The principal wavelength band of the CO_2 laser is 10 600 nm and that of the Nd:YAG laser includes 266 nm, 355 nm, 532 nm and 1 060 nm[2-4].

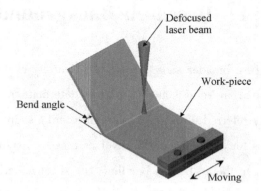

Figure 1.1　Schematic of a straight-line laser forming process

Due to the fact that the laser forming process involves thermal stresses introduced into a metal sheet by laser irradiation rather than external forces, which is employed in the conventional forming methods, the laser forming technique has the following advantages compared to the conventional forming operations[5-10]:

● Manufacturing cost can be reduced because it requires no hard tooling i.e. punch & die system.

● Precise deformation can be achieved because spring-back behavior is not involved which is related to the quality of product.

● A wide variety of complex shaped parts can be obtained through the development of scanning path.

● A wide variety of special material including brittle, hard and thick material can be processed.

● Forming is available in inaccessible areas because it is a non-contact forming process.

Despite its potential advantages, the laser forming technique also has some limitations at the present development stage as follows:

● The forming process is somewhat slow compared to the traditional forming methods of stamp and die.

- The process is energy consuming because of the low energy conversion factor of laser sources.
- The process requires safety protection equipment for the personal because of multidirectional reflection of the laser form the metal.

1.2 Laser Forming Mechanism

Laser forming is a complex transient thermal-mechanical coupling process. The deformation behavior in laser forming process is totally dependent on the temperature field which is determined by the process parameters, material properties and specimen size. The variation of temperature field reduces the different deformation behavior of metal sheet. Therefore, the forming mechanisms are the key point of research on laser forming process. So far, it is not clear how many mechanisms exist in the laser forming. In general, three mechanisms are suggested by many researchers: temperature gradient mechanism (TGM), buckling mechanism (BM) and upsetting mechanism (UM)[12, 14, 27-31]. Recently, two other mechanisms are proposed, which named elastic expansion mechanism (EEM) and coupling mechanism (CM)[32].

1.2.1 Temperature Gradient Mechanism (TGM)

TGM is the most widely known mechanism in laser forming. The deformation relies on steep temperature gradients over irradiated surface. Figure 1.2 shows the temperature gradient mechanism of laser forming[2]. When a laser beam of high power density is rapidly guided across the surface of a metal sheet, the material absorbs a part of the laser energy on the surface and the thermal energy is conducted into the material in lower rate than that of the absorption on the surface, resulting in a steep temperature gradient through the thickness direction as shown in Figure 1.2 (a). As a result, a differential thermal expansion occurs through the thickness direction. Initially, the material expands in the heated zone so that the whole shape of the material bends away from the beam as shown in Figure 1.2 (b). This is called "counter-bending". This thermal expansion is converted into elastic tensile strain and compressive stress because free expansion of the heated material is restricted by surrounding material. Once the stress reaches the temperature gradient flow stress, any additional thermal expansion is converted into a plastic strain. During cooling process, as shown in Figure 1.2(c), the material contracts in the upper layer of the sheet so that the higher tensile stress occurs in the upper region. Therefore, the sheet metal is plastically bent toward the cooling laser path to relax the higher tensile stress. Figure 1.3 illustrates the flow chart of the whole forming process. Usually, the TGM is used to bend thick metal sheet whose thickness is equal or larger than the laser beam spot size.

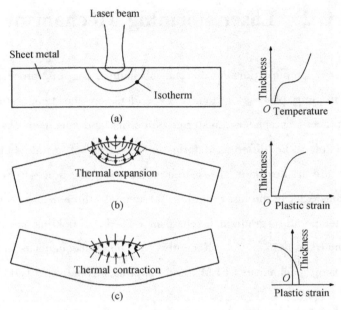

Figure 1.2　Temperature gradient mechanism of laser forming
(a) Laser irradiation; (b) Heating process; (c) Cooling process

1.2.2　Buckling Mechanism (BM)

Even though out-of-plane bending of the sheet by TGM is the most often observed deformation mode, additional deformation modes such as buckling of sheet can also be observed. Compared to the TGM, the BM can be generated by reducing the scanning speed and increasing the laser beam diameter in order to avoid the steep temperature gradient involved in the TGM. When the diameter of the laser beam is significantly larger than (nearly 10 times) the thickness of the metal sheet and when the ratio of the thermal conductivity to the thickness of the material is large, negligibly small temperature gradient in thickness direction is also favored where the ratio of the thermal conductivity to the thickness of the material is large. Unlike the TGM, the bending direction in the BM is not changing during the heating and cooling process. An important feature of the BM is that the direction of the geometry change is not defined by the process itself. The buckle can be towards or away from the laser beam. It depends on the boundary conditions, mainly to the pre-curvature of the metal sheet, internal stresses and external or gravitational forces. Therefore, the BM may be employed for bending thin sheets along a straight line towards or away from the laser beam. It is also proposed to be used for a tube bend.

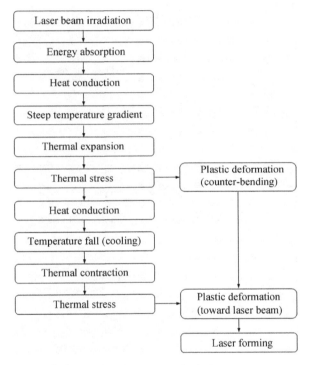

Figure 1.3 Flow chart of laser forming process

Figure 1.4 shows the developing stages of the BM[2]. Figure 1.4 (a) illustrates the compressive thermal stresses generated by a laser beam. It can be seen from the figure that there is no steep temperature gradient in the metal sheet. Figure 1.4 (b) shows buckling starting to develop in the metal sheet which is originated from instability dueto thermal stresses. Figure 1.4 (c) presents the development of both the plastic deformation and the elastic deformations in the metal sheet. The plastic buckling occurs preferentially at the top of the metal sheet because the flow stress is low in this region due to the temperature rise, and the elastic deformation is involved at the neighboring region of the plastic buckling due to the lower heating. Forces caused by the elastic deformation are counteracted by the constraints from surrounding material. The buckling increases along the scanning line. Figure 1.4 (d) show the full development of a bend angle. The buckle is generated across the whole metal sheet when the laser beam leaves the sheet surface because no restraining forces against the elastic strains are left, and the elastic strain relax becoming straight, whereas the plastic bend remains.

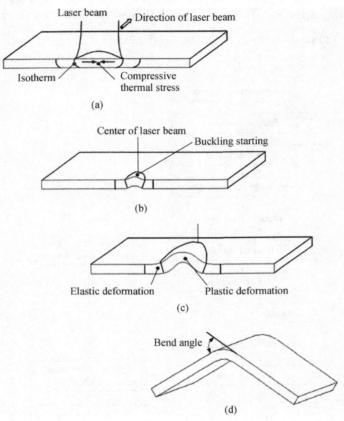

Figure 1.4　Buckling mechanism of laser forming

(a) generation of thermal compressive stresses by laser irradiation; (b) development of buckling by thermal stresses;
(c) development of elastic and plastic deformations; (d) development of a bend angle

1.2.3　Upsetting Mechanism (UM)

Similar to the BM, the UM of laser forming is operative when the temperature gradient in the laser heated region is negligibly small. However, the thickness of the sheet for UM is getting much larger and the width is getting smaller than the original material, as shown in Figure 1.5[2]. The localized heating of the material by the laser irradiation causes uniform thermal expansion of the material throughout the thickness of the laser-heated region. The geometry of the sheet prevents the buckling of the sheet. Instead, local shortening of the sheet in the laser-heated region takes place resulting in the local increase in thickness of the sheet. This mechanism can be used to form a metal sheet into a specially formed part with a proper heating strategy. Aligning, adjustment and rapid prototyping are also possible.

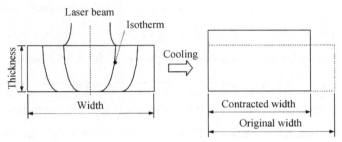

Figure 1.5 Upsetting mechanism of laser forming

1.2.4 Elastic Expansion Mechanism (EEM)

When a point or a small zone of the metal sheet is irradiated by a laser beam, as shown in Figure 1.6[37], a larger thermal expansion occurs at the heated zone than that in the TGM. Due to the significantly large expansion capacity, the plastic compressive stress from the surrounding cool material is removed but only leaving the elastic compressive stress resulting from the elastic expansion. This elastic compressive stress induces an elastic deformation so that a small bending occurs in the metal sheet. If a larger deformation is desired, it is efficient to irradiate the points or small zones adjacent to the heated zone but not heating the same position repeat because the last elastic deformation will be removed by the current irradiation. Aligning and adjustment is available using the EEM.

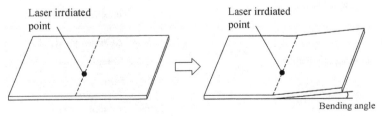

Figure 1.6 Elastic expansion mechanism of laser forming

1.2.5 Coupling Mechanism (CM)

Coupling mechanism, as the name suggests, is the coupling of two kinds of mechanisms which include the TGM and UM[37]. As the previous description, in the TGM, the plastic compressive deformation occurs only at the top surface of the metal sheet and there is no plastic deformation at the middle surface. The sheet bends along the laser scanning path in the TGM. However, in the UM, shortening occurs in the whole metal sheet because the plastic compressive deformations at the top surface and bottom surface are almost the same. In the CM, the plastic deformation occurs at the middle surface of the metal sheet and the deformation at the top surface is larger than that at the middle surface. Therefore, there is not only bending deformation but also shortening deformation in the CM. Thus, the CM is also proposed to be used for aligning and adjustment.

As stated above, in the laser forming process, a forming mechanism is mainlydetermined by the temperature distributions produced inside the material. Thus, a desired mechanism can be generated by controlling laser parameters (i.e., laser power, scanning speed and laser beam diameter) and specimen dimensions. The influences of laser parameters and specimen dimensions on final bending angle are also the main research objectives of this paper. The five laser forming mechanisms described above are summarized from principle, forming result, process parameter and applications in Table 1.1.

Table 1.1 Summary of laser forming mechanisms

Mechanism	Principle	Forming Result	Process Parameter	Applications
TGM	Temperature gradient mechanism	Towards laser beam	Rapid scanning speed Small beam diameter Thick sheet	Thick sheet bending
BM	Buckling mechanism	Depending on boundary conditions	Slow scanning speed, Large beam diameter, Thin sheet	Thin sheet bending, Tube bending
UM	Upsetting mechanism	Thickness increase	Slow scanning speed, Small beam diameter, Thick and stiff sheet	Aligning Adjustment

续表

Mechanism	Principle	Forming Result	Process Parameter	Applications
EEM	Elastic expansion mechanism	Towards laser beam	Small zone irradiation, Smalll beam diameter, Thick sheet	Aligning Adjustment
CM	Coupling mechanism	Towards laser beam + thickness increase	Small beam diameter, Thick sheet	Aligning Adjustment

1.3 Research Trends and Applications

The research methodologies for the laser forming over about three decades can be classified into three categories-experimental study, analytical modeling and numerical simulation[2]. The experimental study, as the focus of earlier researches on laser forming, has concentrated on investigating material types amenable to laser forming, metallurgical variations after forming and scanning strategies of laser forming. Edwardson et al.[9] studied different scanning strategies to develop a saddle shape from a flat sheet. Ojedaa and Grez[11] studied a few of other scanning strategies, like zigzag, squared and stepped scan lines for stainless steel thin sheets.

Analytical models, which emerged mostly in the 1990s, have contributed to characterizing laser forming mechanisms and estimating bending angle. Vollertsen[12] suggested that three kinds of mechanisms exist in laser forming, as described previously, temperature gradient mechanism (TGM), buckling mechanism (BM) and upsetting mechanism (UM). Vollertsen and Rodle[13] proposed an analytical model to study the temperature gradient mechanism of laser bending. Arnet and Vollertsen[14] found that laser forming can not only be used to generate concave shapes but also to generate convex shapes of a low thickness, with materials with high heat conduction, by a beam of wide diameter, by a relatively high laser power and by a low path feed rate. Geiger and Vollertsen[12, 15] proposed a two-layer model based on the temperature gradient mechanism with the temperature field obtained by an energy approach or by solving the 3D heat conduction problem. Cheng and Lin[16-17] established an analytical model to describe the 3D temperature field with a Gaussian heat source and estimated the bending angles.

As the development of numerical analysis software, finite element simulations have been carried out to overcome the limitations of the experimental study and analytical approach and to improve the fundamental knowledge of laser forming. Finite element simulations have concentrated on investigating the temperature distribution, stress and strain distributions during laser scanning and effects of laser parameters and specimen size on final bending angles. Ji and Wu[18] investigated the temperature distribution through laser forming simulation. Li and Chen[19] analyzed not only the stress and strain distributions of pure aluminum plate with different thickness, but also the influence of scanning speed on temperature fields and plastic strain of metal sheet. Cheng et al.[20] investigated the effects of sheet length and width on final bending angle through both experiment and FEM simulations.

Recently, researches on laser forming have been extended to various engineering applications. New trends include the laser forming of complex or irregular shapes[9], laser forming of tubes[21], laser forming with preload[22], laser forming of high strength material[23], laser forming of composite material[24-25] and laser forming with two beams[26].

The numerous applications of laser forming process in use today are considered cost effective in the design and manufacturing process. These applications are shown in Figure 1.7, which include:
- Plate bending for complex shapes.
- Tube and pipe precision forming.
- Unbending techniques for repairs and alignment applications.
- Thermal pre-stressing.

• Plate bending

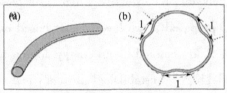

• Tube and pipe precision forming

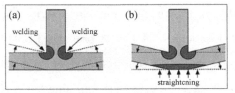

- Unbending process for repairs and alignment

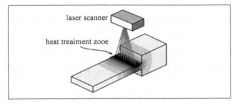

- Thermal pre-stressing

Figure 1.7 Applications of the laser forming process

1.4 Research Motivation and Objectives

Researches to-date on laser forming, as referred previously, have mostly focused on easily-deformed materials, such as mild steel, aluminum, titanium, and so on. However, with the development of industry, especially in automotive industry and aerospace field, these easily-deformed materials cannot meet the increasing demands any more. Light weight material is required for saving the fuel consumption while high strength is essential for the safety at the same time. In this case, a series of high strength steels, for instance, DP980 (Dual Phase 980 steel) and CP1180 (Complex Phase 1180 steel) are getting more and more popular in automotive industry. Unfortunately, these materials are much more difficult to be deformed than other metal. Thus, it is important to find a forming process suitable for the difficult deformed materials. Laser forming is one of the processes to bend the high strength steels easily and precisely.

Previous researches on laser forming have mostly focused on experimental method and analytical method, which are cost consuming and time consuming and have limitations for expanding results into the general case. In addition, the forming mechanism was not yet fully understood by experimental method and analytical method because the laser induced forming process involves very complex thermal deformations influenced by process parameters such as laser power, scanning speed, scanning direction, and material properties and so on. In order to describe the laser forming behavior during the process and characterize the quality of formed parts, theFEM simulations have great advantages. A few of previous FEM simulations on laser forming process have used the simple square heat source model or small stroke jumping heat source model[23, 41-42]. However, these heat source models can simulate neither the real laser spot nor the real heat transfer process. In fact, the laser heat source employed in laser forming obeys Gaussian distribution, which should be established in the FEM simulations aiming to obtain a more precise result.

Thus, the objective of this research is focused on the forming characteristics of DP980 high strength steel using Gaussian heat source by finite element analysis. The main contents include:

● Analysis of laser parameters effect, i. e. laser power, laser scanning speed and line energy.

● Analysis of specimen size effect, i. e. sheet length, width and thickness.

● Analysis of scanning pattern effect, i. e. scanning directions in two-time scanning and three-time scanning process.

1.5 Outline of the Paper

The organization of the work in this paper is outlined as follows:

Chapter 2 gives a review of the finite element formulations for both the heat transfer and the structural problems which are involved in the laser forming.

Chapter 3 deals with the finite element modeling for the thermal and coupled temp-displacement deformation in the simulation and compares the simulation results with the experimental results to confirm the veracity of finite element analysis.

Chapter 4 illustrates the finite element analysis of one-time laser scanning process, in which the effects of laser parameters and specimen size on the forming characteristics of DP980 high strength steel are described in detail.

Chapter 5 extends the finite element analysis to multi-time laser scanning process. The effects of scanning patterns on forming characteristics of DP980 high strength steel are investigated.

Finally, Chapter 6 summarizes the conclusions and suggests some recommendations for the future work.

2 Theory of Finite Element Analysis

2.1 Introduction

The aim of this chapter is to review the theoretical background of the finite element method (FEM) employed in the analysis of the laser forming processes. Both the thermal problem and structural problem, which are involved in the laser forming process, can be analyzed using the finite element method. The thermal problem includes the heating of the laser beam on the metal surface and the heat transfer inside the material. The structural problem involves the plastic stress converted from the thermal stress and the final plastic deformation of the material.

2.2 Finite Element Method (FEM)

FEM is a computer-based numerical method that is employed to analyze the laser forming process. In FEM, the bodies of user-defined geometry arediscreted into a large amount of elements. Each element has nodes that are used to represent the value of the field variable, for example, temperature, stress and displacement, over the element by interpolation function. The matrix equations expressing the properties of the individual elements are determined using one of the three approaches—the direct approach, the variational approach, or the weighted residuals approach[2]. The individual matrix equations are assembled to find the overall system equations.

The concepts of FEM are briefly presented in terms of continuum problems, the weighted residuals approach and interpolation functions. In the continuum problem, all bodies are taken into account continuous at all points in space and all field quantities describing the state of the body are sufficiently differentiable in the independent variables of the continuum, space and time. Continuum problems are concerned with the fields of temperature, stress, mass concentration, displacement and so on. They can be approximately represented by partial differential equations and their boundary conditions. The governing equations are often called boundary value problems because their solution is determined in some domain defined by a given boundary on which boundary conditions are specified and can be formulated in general terms to be solved. The weighted residuals approach is a global technique for obtaining approximate functional representation of the governing linear and nonlinear partial differential equations in a way as to formulate the element equations. Interpolation functions are used to express the behavior of a field variable within an element. They involve quantities at any point in the interior of the element to the nodal point values. The interpolation functions have to provide continuous functions for the interpolated field within an element. Usually, engineering problems addressed by the FEM include structural, thermal, fluid flow, electrical, magnetic, acoustic noise and so on. Both thermal and structural problems are involved in the laser forming process[45-46].

2.3 Thermal Problem

The laser forming process is achieved by heating of a laser beam on the surface of a metal sheet. The heat transfer modes in the problem are characterized by conduction, convection and radiation. Part of the heat is transferred inside the plate by conduction and part of the heat is transferred to the surrounding material by convection and radiation effects, as shown in Figure 2.1[2]. The temperature distributions in the metal sheet can be calculated using the finite element method according to the heat transfer equations.

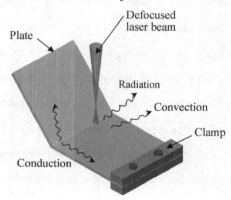

Figure 2.1 Heat transfer modes involved in the laser forming process

2.3.1　Heat Diffusion Equation

Considering a homogeneous medium and the temperature distribution $T(x, y, z)$ is expressed in Cartesian coordinates. The temperature distribution in a differential control volume defined in the medium is governed by the law of conservation of thermal energy as follows[50]:

$$\rho c_p \frac{\partial T}{\partial t} = \frac{\partial}{\partial x}\left(k \frac{\partial T}{\partial x}\right) + \frac{\partial}{\partial y}\left(k \frac{\partial T}{\partial y}\right) + \frac{\partial}{\partial z}\left(k \frac{\partial T}{\partial z}\right) + q \quad (2.1)$$

where, ρ—density of the medium;

c_p—specific heat of the medium;

k—thermal conductivity of the medium;

$\dot{q}$—rate at which heat is generated per unit volume of the medium;

$\rho c_p \dfrac{\partial T}{\partial t}$— time rate of the thermal energy variation of the medium per unit volume;

$k \dfrac{\partial T}{\partial x}, k \dfrac{\partial T}{\partial y}$ and $k \dfrac{\partial T}{\partial z}$—rate of heat transfer per unit area (heat flux) at the x, y and z coordinate locations, respectively.

Eq. (2.1) is the general form of heat transfer equation. Using this equation, the temperature distribution $T(x, y, z)$ is obtained as a function of time.

For heat conduction from one surface of a metal sheet to another, the heat flux equation is more generally stated by Fourier's Law[2]:

$$q'' = -k\left(k \frac{\partial T}{\partial x} + \frac{\partial T}{\partial y} + \frac{\partial T}{\partial z}\right) = -k \nabla T \quad (2.2)$$

The negative sigh means that heat is transferred in the direction of decreasing temperature.

Thus, Eq. (2.1) can be simply expressed:

$$\rho c_p \frac{\partial T}{\partial t} = -\nabla \cdot q'' + \dot{q} \quad (2.3)$$

In order to determine the temperature distribution in the medium, Eq. (2.3) need to be solved at the given boundary conditions. The boundary conditions at defined surfaces encountered in heat transfer are presented as follows:

For specified surface temperature, T_s,

$$T_{(x,y,z,t)} = T_s \quad (2.4)$$

For specified surface heat flux, q'',

$$-k\nabla T = q'', \text{ for finite heat flux} \qquad (2.5)$$

$$\nabla T = 0, \text{ on adiabatic surface} \qquad (2.6)$$

For convection surface condition,

$$-k\nabla T = h[T_\infty - T_{(x,y,z,t)}] \qquad (2.7)$$

where, h—a convective heat transfer coefficient;

T_∞—the surrounding temperature.

For radiation heat exchange,

$$-k\nabla T = \alpha q_r'' - \sigma\varepsilon T^4_{(x,y,z,t)} \qquad (2.8)$$

where, α—surface absorptivity;

q_r—incident radiant heat flux;

σ—Stefan-Bolzmann constant;

ε—surface emissivity, which is a function of the surface temperature.

It is found that the heat transfer problem is nonlinear due to the temperature-dependent material properties and the nonlinear boundary conditions.

2.3.2 Finite Element Formulation

A general formulation of element equations for transient heat transfer in a medium can be derived from a 3D solid Ω bounded by a surface Γ with various heat transfer modes as shown in Figure 2.2[45].

Assuming the solution Ω is divided into M elements and each element has r nodes, then the temperature and temperature gradients can be expresses as follows:

$$T^{(e)}(x,y,z,t) = \sum_{i=1}^{r} N_i(x,y,z)T_i(t) \qquad (2.9)$$

$$\frac{\partial T^{(e)}}{\partial x}(x,y,z,t) = \sum_{i=1}^{r} \frac{\partial N_i}{\partial x}(x,y,z)T_i(t) \qquad (2.10)$$

$$\frac{\partial T^{(e)}}{\partial y}(x,y,z,t) = \sum_{i=1}^{r} \frac{\partial N_i}{\partial y}(x,y,z)T_i(t) \qquad (2.11)$$

$$\frac{\partial T^{(e)}}{\partial z}(x,y,z,t) = \sum_{i=1}^{r} \frac{\partial N_i}{\partial z}(x,y,z)T_i(t) \qquad (2.12)$$

where, N_i—the temperature interpolation function;

$T_i(t)$—the value of the temperature at each node.

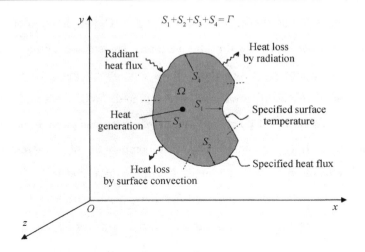

Figure 2.2 Heat transfer modes involved in a 3D solid Ω

Eq.s (2.9) ~ (2.12) can be simply expressed in matrix form:

$$T^{(t)}(x,y,z,t) = N(x,y,z)T(t) \tag{2.13}$$

$$\nabla T_{(x,y,z,t)} = B_{(x,y,z)}T(t) \tag{2.14}$$

where, $T(t)$——the vector of element nodal temperatures;

$N_{(x,y,z)} = [N_1\ N_2 \cdots N_r]$——the temperature interpolation matrix;

$$B_{(x,y,z)} = \begin{bmatrix} \dfrac{\partial N_1}{\partial x} & \dfrac{\partial N_2}{\partial x} & \cdots & \dfrac{\partial N_r}{\partial x} \\ \dfrac{\partial N_1}{\partial y} & \dfrac{\partial N_2}{\partial y} & \cdots & \dfrac{\partial N_r}{\partial y} \\ \dfrac{\partial N_1}{\partial z} & \dfrac{\partial N_2}{\partial z} & \cdots & \dfrac{\partial N_r}{\partial z} \end{bmatrix}$$ ——the temperature gradient interpolation matrix.

2.4 Structural Problem

As described above, the thermal expansion is caused by heating of the laser beam. This thermal expansion is enough to produce large plastic strains in the metal sheet via a mechanical process.

2.4.1 Stress Equation

Consider a three dimensional body in equilibrium under the action of a set of the external forces (F_1, F_2, F_3 and F_4), as shown in Figure 2.3 (a). The application of the external forces results in the development of stresses within the medium. The stress, acting on a small area, can be divided into the normal stress and two shear stress components and. Thus, if the three values are known, the stress on the small area can be calculated. When a small cube is taken into account, as shown in Figure 2.3 (b), nine values of stress components are necessary for the stresses in the body to be fully calculated. The nine components of the stress are known as the stress tensor, which can be displayed as elements of the square matrix[47, 51]:

$$\boldsymbol{\sigma} = \begin{bmatrix} \sigma_{11} & \sigma_{12} & \sigma_{13} \\ \sigma_{21} & \sigma_{22} & \sigma_{23} \\ \sigma_{31} & \sigma_{32} & \sigma_{33} \end{bmatrix} = \begin{bmatrix} \sigma_{xx} & \tau_{xy} & \tau_{xz} \\ \tau_{yx} & \sigma_{yy} & \tau_{yz} \\ \tau_{zx} & \tau_{zy} & \sigma_{zz} \end{bmatrix} \quad (2.15)$$

In indicial notation, the stress matrix is given by

$$\sigma = \sigma_{ij}, \ i = x, y, z, \ j = x, y, z \quad (2.16)$$

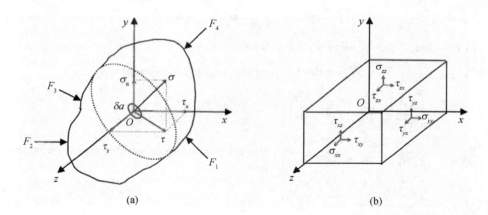

Figure 2.3 Stress acting on a small area

(a) Normal and shear stresses; (b) Components of stress tensor

2.4.2 Finite Element Formulation

Consider a body in equilibrium occupying a finite volume V and bounded by a surface S, which is subjected to external tractions $T_j (j = x, y, z)$ acting on a point P, as shown in Figure 2.4. The equilibrium equation can be expressed as follows[51–52]:

$$\int_S \sigma_{ij} n_i \mathrm{d}S + \int_V \rho g_j \mathrm{d}V = 0 \quad (2.17)$$

where, σ_{ij}—the stress tensor;

n_i—the outward unit normal to the tractions, $T_j(j=x,y,z)$;

ρ—the density of the body;

g_j—the body force per unit mass $(j=x,y,z)$.

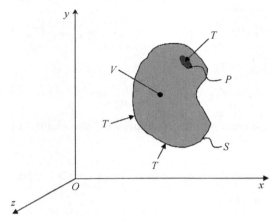

Figure 2.4 External tractions acting on a point P

Using Green's theorem,

$$\int_V \left(\frac{\partial \sigma_{ij}}{\partial x_i} + \rho g_j \right) dV \tag{2.18}$$

Then, the equilibrium condition becomes

$$\frac{\partial \sigma_{ij}}{\partial x_i} + \rho g_j = 0 \tag{2.19}$$

At the boundary of the body, the force T_i must be equal to the stress vectors:

$$\left. \begin{array}{l} T_x = n_x \sigma_{xx} + n_y \sigma_{yz} + n_z \sigma_{zx} \\ T_y = n_x \sigma_{xy} + n_y \sigma_{yy} + n_z \sigma_{zy} \\ T_z = n_x \sigma_{xz} + n_y \sigma_{yz} + n_z \sigma_{zz} \end{array} \right\} \tag{2.20}$$

Eq. (2.20) can be written in indicial notation:

$$T_j = n_i \sigma_{ij} \tag{2.21}$$

To determine the stress components uniquely, boundary conditions of displacement are necessary.

Considering a continuous velocity field v_j, the rate of work done by the distribution of surface traction T_i is,

$$\int_S T_i v_j dS = \int_S n_i \sigma_{ij} v_i dS = \int_V \frac{\partial}{\partial x_i}(\sigma_{ij} v_j) dV = \int_V \sigma_{ij} \frac{\partial v_i}{\partial x_i} dV \tag{2.22}$$

Applying Green's theorem to the vector $\sigma_{ij} v_j$, the equilibrium equation is obtained in an integral form:

$$\int_S T_i v_j \mathrm{d}S = \int_V \sigma_{ij}\varepsilon_{ij}\mathrm{d}V \qquad (2.23)$$

where, ε_{ij}—the displacement rate.

The variational principle for the equilibrium equation, Eq. (2.23), is expressed according to Chakrabarty[52] and Huang[53]:

$$\delta\int\left[\dot{\sigma}_{ij}\dot{\varepsilon}_{ij} - \sigma_{ij}\left(2\dot{\varepsilon}_{ik}\dot{\varepsilon}_{jk} - \frac{\partial v_k}{\partial x_i}\frac{\partial v_k}{\partial x_j}\right)\right]\mathrm{d}V - 2\int \dot{T}_j v_j \mathrm{d}S = 0 \qquad (2.24)$$

Here, the second term in the volume integral represents the change in geometry due to the large strain.

For a finite volume element, the non-zero components of the Jaumann stress rate are described by a column vector $\dot{\boldsymbol{\sigma}}$:

$$\dot{\boldsymbol{\sigma}} = \boldsymbol{C}\dot{\boldsymbol{\varepsilon}} \qquad (2.25)$$

where, $\boldsymbol{C}$— the constitutive matrix.

In the variational form expressed by Eq. (2.24), the second term in the volume integral can be omitted because the positional changes are negligible when the strains are small. Thus, the matrix form of this equation becomes:

$$\delta\left(\int \dot{\boldsymbol{\varepsilon}}^\mathrm{T} \boldsymbol{C}\dot{\boldsymbol{\varepsilon}}\mathrm{d}V - 2\int \boldsymbol{v}^\mathrm{T}\boldsymbol{T}\mathrm{d}S\right) = 0 \qquad (2.26)$$

The velocity distribution in a typical element and the associated strain rate can be expressed in the general form:

$$\boldsymbol{v} = \boldsymbol{N}\boldsymbol{d} \qquad (2.27)$$

$$\dot{\boldsymbol{\varepsilon}} = \boldsymbol{B}\boldsymbol{d} \qquad (2.28)$$

where, $\boldsymbol{d}$— the displacement vector;

$\boldsymbol{N}$ and $\boldsymbol{B}$—the displacement and displacement rate interpolation matrix, respectively.

Substituting Eqs. (2.27) and (2.28) into Eq. (2.26), the variational principle is reduced to

$$(\delta\boldsymbol{d})^\mathrm{T}\left(\int \boldsymbol{B}^\mathrm{T}\boldsymbol{C}\boldsymbol{B}\boldsymbol{d}\mathrm{d}V - \int \boldsymbol{N}^\mathrm{T}\dot{\boldsymbol{T}}\mathrm{d}S\right) = 0 \qquad (2.29)$$

Since the variation of the nodal velocities is arbitrary, the above equation requires the expression in the parenpaper to vanish. Therefore, the rate equilibrium equation for the element becomes

$$\boldsymbol{k}\boldsymbol{d} = \dot{\boldsymbol{r}} \qquad (2.30)$$

where, $\boldsymbol{k} = \int \boldsymbol{B}^\mathrm{T}\boldsymbol{C}\boldsymbol{B}\mathrm{d}V$—the element stiffness matrix;

$\dot{r} = \int N^{\mathrm{T}} \dot{T} \mathrm{d}S$—the associated load rate vector.

In problems involving large plastic strain, the variational form leads to an additional stiffness term in the finite element equation. To derive the second stiffness matrix,

$$v = N_i d \qquad (2.31)$$

$$\dot{\varepsilon}_{ij} = B_{ij} d \qquad (2.32)$$

where, N_i and B_{ij}—the row vectors identical to the rows of N and B, respectively.

Then, the rate equilibrium equation, Eq. (2.30) becomes

$$(k + k_s) d = \dot{r} \qquad (2.33)$$

where, $k_s = \int (N_{k,i}^{\mathrm{T}} \sigma_{ij} N_{k,j} - 2 B_{ki}^{\mathrm{T}} \sigma_{ij} B_{kj}) \mathrm{d}V$.

Therefore, the global equilibrium equation for elastic-plastic problems is expressed:

$$KU = \dot{R} \qquad (2.34)$$

where, K—the global stiffness matrix;

U—the global displacement vector;

$\dot{R}$—the global load rate vector.

2.5 Summary

This chapter has given a brief description of concept of the FEM and the equation derivations of the finite element equations involved in the thermal problem and structural problem. The temperature distribution in a differential control volume is governed by the law of conservation of thermal energy and the heat conduction from one surface of a metal sheet to another is illustrated by Fourier's Law. In the structural problem, the stress is calculated by the stress tensor which can be displayed as elements of a square matrix. The matrix equation involving the large plastic strain is formulated from the equilibrium equation.

3 Finite Element Modeling and Experimental Verification of Laser Forming Process

3.1 Introduction

In this chapter, a finite element modeling of DP980 steel sheet is established to perform the simulation of laser forming process using the commercial finite element software ABAQUS[38]. DP980 is one of the kinds of high strength steel that is gaining more and more favor in the automotive industry in recent years due to its high yield strength and light weight. The chemical composition of DP980 steel is presented in Table 3.1. In the simulation, a nonlinear, transient, three-dimensional heat transfer model and a nonlinear, three-dimensional structural model are developed to illustrate both the thermal and mechanical responses of the material during laser forming process. The Gaussian laser source model is defined using ABAQUS user subroutine DFLUX. In order to verify the validity of the finite element analysis, an experimental work for the same material at the same laser scanning conditions with the simulation is implemented.

Table 3.1 Chemical composition of DP980 steel

Material	Composition/(%)					
	C	Mn	P	S	Si	Fe
DP980 steel	0.065	2.40	0.023	0.004	0.076	Bal.

3.2 3D Finite Element Modeling

FEM analysis is an effective method on researching the deformation behavior of the material in forming process. It can shorten the design cycle and improve the quality of the products. In this study, the commercial finite element software ABAQUS v6.10 is employed and a few of assumptions are made for the FEM analyses:

- The initial temperature of the material is room temperature.
- The metal sheet is an isotropic material and its physical properties (yield strength, Young's modulus, Poisson's ratio, specific heat) vary with the temperature.
- The expansion and absorption coefficient of the material do not vary with thetemperature.
- Air convection and radiation are taken into account.
- The yield of the material obeys VonMises yield criterion[47-49].
- Heat generated by plastic deformation is negligible compared with the heat input by laser energy.
- Melting is not involved in the metal sheet during the forming process.
- Phase transformation is not accounted for in the simulation.
- The influence of gravity is neglected.

3.2.1 3D Part and Material Properties

The FEM analysis of the laser forming process is a transient coupled temp-displacement process. The schematic of the laser forming process involved in this study is illustrated in Figure 3.1. A rectangular DP980 steel sheet is fixed by a clamp at one end side and the other end side is left free. It is heated along its centerline from point A to point B by the irradiation of a laser beam. The specimen dimensions and laser parameters used in the current simulation are listed in Table 3.2.

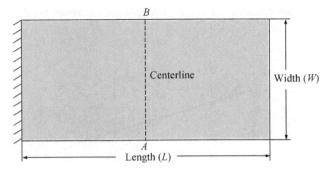

Figure 3.1 Schematic of laser forming process

Table 3.2 Parameters and values involved in laser forming process

Specimen dimensions			Laser parameters		
Length mm	Width mm	Thickness mm	Power W	Scanning speed (m·min^{-1})	Beam diameter mm
100	50	1.4	1 000	2	5.2

The material properties related to the deformation problem in the simulations include: density, stress-strain curve, Young's modulus, Poisson's ratio, thermal conductivity and thermal expansion coefficient. Density of the material is $7.8 \times 10^3 \text{kg/m}^3$ and coefficient of thermal expansion is 15.2×10^{-6} ℃. The stress-strain curve, Young's modulus, Poisson's ratio and thermal conductivity dependent on the temperature are shown in Figure 3.2 ~ Figure 3.6, respectively.

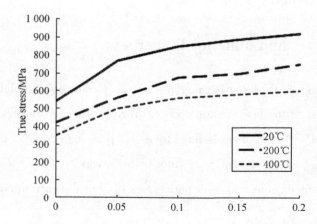

Figure 3.2 Stress-strain curves against temperature of DP980 high strength steel

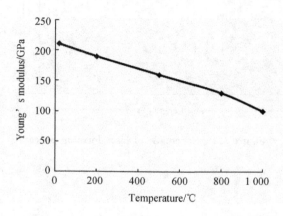

Figure 3.3 Young's modulus against temperature of DP980 high strength steel

3 Finite Element Modeling and Experimental Verification of Laser Forming Process

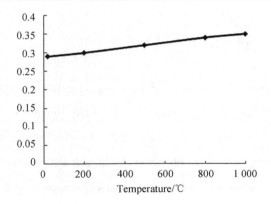

Figure 3.4 Poisson's ratio against temperature of DP980 high strength steel

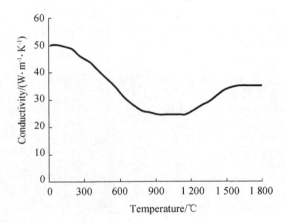

Figure 3.5 Thermal conductivity against temperature of DP980 high strength steel

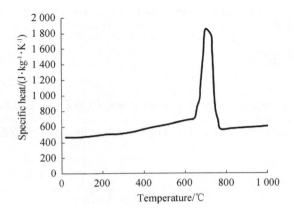

Figure 3.6 Specific heat against temperature of DP980 high strength steel

3.2.2 Element Selection

Due to the small thickness of the steel sheet comparing to its length, a shell model is used for the three dimensional modeling of the sheet, and the shell model is defined as the top surface of the sheet. There are five integration points from top to bottom surface in the thickness direction, at which the temperature, stress and strain are calculated. A 4-node thermal shell element type S4RT is selected for the finite element analyses and the structured meshing technique is used for the accurate results owing to the regular shape of the sheet. In order to shorten the running time of the computer, a local mesh refinement technique is employed nearby the laser path, as shown in Figure 3.7. The output data available from the analysis results include stress and strain, node temperature and node displacement, which will be illustrated in detail below.

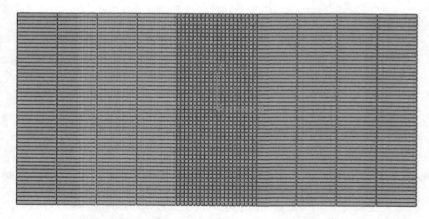

Figure 3.7 Local mesh refinement technique for 3D metal sheet part

3.2.3 Heat Source

The laser heat source employed in the simulations obeys Gaussian distribution, as shown in Figure 3.8, which is defined using ABAQUS user subroutine DFLUX. The heat flux is described as follows[39]:

$$I = \frac{2AP}{\pi R^2}\exp\left(-\frac{2r^2}{R^2}\right) \tag{3.1}$$

where, I—the intensity of the incident heat flux at the material surface;

A—the absorptivity of laser energy;

P—laser power;

R—laser beam diameter;

r—distance from the center of the laser beam.

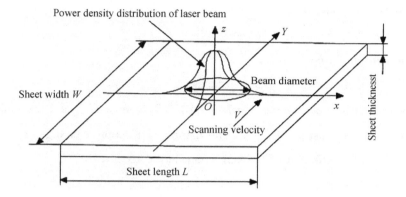

Figure 3.8 Gaussian-distributed laser heat source model

The mean thermal flux density, $\overline{A}$, within the area of the laser beam is calculated from Eq. (3.1) as follows:

$$\begin{aligned}\overline{A} &= \frac{1}{\pi R^2}\int_0^R I \times 2\pi r dr \\ &= \frac{2\pi}{\pi R^2}\int_0^R \frac{2AP}{\pi R^2}\exp\left(-\frac{2r^2}{R^2}\right)rdr \\ &= \frac{0.865AP}{\pi R^2}\end{aligned} \quad (3.2)$$

In the present simulation, the heat flux input and the movement of laser beam arerealized by a user subroutine DFLUX of ABAQUS. The code is written using Microsoft Visual Studio 2008. The heat flux obtained from Eq. (3.2) is used as a surface load and the load is imposed on the top surface of the DP980 steel sheet.

3.2.4 Boundary Conditions

Boundary conditions involved in the heat transfer model mainly indicate the heat losses from the sheet surfaces to the surroundings due to the natural convection and radiation effects.

For the radiant heat loss,

$$q_r = \sigma\varepsilon(T_r^4 - T_\infty^4) \quad (3.3)$$

where, σ—Stefan-Boltzmann constant;

ε—the emissivity of radiating surface;

T_r—the absolute temperature of radiating surface;

T_∞—the absolute temperature of surrounding.

For the convective heat loss,

$$q_c = h_f(T_S - T_\infty) \tag{3.4}$$

where, h_f—the convection coefficient;

T_∞—the temperature at the plate surfaces.

The Stefan-Blotzmann constant is 5.6703×10^{-8} W/($m^2 \cdot K^4$) and the emissivity of the radiating surface is kept constant of 0.85. The natural convection exchange coefficient is taken as 5 W/($m^2 \cdot K$) and coefficient of thermal expansion is 15.2×10^{-6} ℃.

3.2.5 Finite Element Analysis Result

The simulation result is shown in Figure 3.9 in which the contour shows the stresscontribution and the deformation is magnified 10 times for seen clarity. It is seen that the DP980 steel sheet bent towards the laser beam. The temperature distribution of the center point of the sheet and the final bending angles along the width direction are exported from the analysis result, as shown in Figure 3.10 and Figure 3.11. In Figure 3.11, the normalized length is defined as a ratio X/L, where X means the distance from the measuring point to the laser starting point (point A in Figure 3.1) and L is the length of the whole laser scanning path (the width of the sheet in the current simulations). It is found from Figure 3.11 that the average bending angle along the width of the sheet is $1.27°$.

Figure 3.9 FEM simulation of the laser forming process

(Deformation scale factor = 10)

3 Finite Element Modeling and Experimental Verification of Laser Forming Process

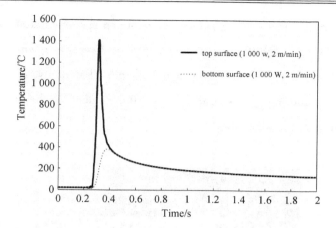

Figure 3.10 Temperature histories at the center point of the steel sheet
($P = 1\ 000$ W, $v = 2$ m/min)

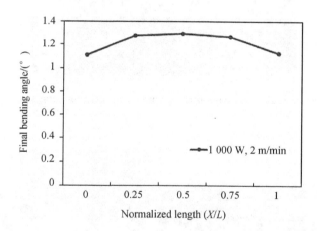

Figure 3.11 Final bending angles along the width direction of the steel sheet
($P = 1\ 000$ W, $v = 2$ m/min)

3.3 Experimental Verification

In order to verify the result of the FEM analysis above, an experiment for the same specimen is complemented under the same conditions. The laser system employed in the experiments is a PRC-1500 CO_2 laser, with a maximum output power of 1.5 kW, as shown in Figure 3.12. In the experiment, the specimen is clamped by the fixture of the working table and the laser scanning process is actually realized by moving the working table against the fixed laser beam using a given speed. Figure 3.13 shows the deformed specimen after the experiment using the laser system. The average final bending angle of five specimens, measured using a protractor, is 1.23° which is in good agreement with the FEM analysis result (1.27°). The small difference is considering to be resulted from the neglect of the heat transfer from the specimen to the fixture in the simulation.

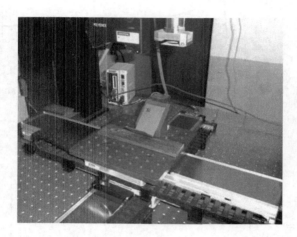

Figure 3.12   Experimental equipment of laser forming system

(a)   (b)

Figure 3.13   A DP980 steel sheet deformed by laser forming system

(a) top view; (b) front view

3.4 Summary

In this chapter, the commercial finite element software ABAQUS v 6.10 is employed to design the 3D specimen model and Gaussian heat source model in order to perform the FEM analysis of laser forming process. In order to verify the result of the FEM analysis, an experiment under the same conditions is complemented and the result shows that a good agreement exists between the analysis result and the experiment result. It is concluded that the FEM simulation is valid for the analysis of laser forming process and it is a powerful tool to predict the deformation and investigate the effects of laser parameters, specimen dimensions and scanning patterns on the forming characteristics.

4 Finite Element Analysis of One-time Laser Scanning

4.1 Introduction

The forming behavior in a laser forming process is achieved by heating of the laser beam without any external force. Due to it is a complex coupled temp-displacement problem, the forming behavior depends on the temperature distributions inside the material, which is mainly determined by the laser parameters, such as laser power, laser scanning speed and line energy. Furthermore, due to the plastic deformation is induced by the temperature gradient in the thickness direction of the material, the specimen size, i.e. length, width and thickness, also plays an important role in the laser forming process. In this chapter, the effects of the laser parameters and specimen sizes on forming characteristics in one-time laser forming process are investigated by the finite element analyses.

4.2 Analysis of Laser Parameter Effect

4.2.1 Effect of Laser Power

As one of the important laser parameters, laser power affects the forming characteristics significantly. In order to investigate the effect of laser power, another simulation under the same conditions with the previous simulation but only changingthe laser power from 1 000 W to 750 W, is carried out. Figure 4.1 shows the temperature comparisons at the center point of the sheet in the two cases of simulations. It is seen that when the laser power is decreased to 750 W, the temperatures on both the top surface and the bottom surface are lower than those of 1 000 W. It means that the temperature decreases with the decrease of the laser power. That is because the heat flux flowing into the material at the unit time decreases when the laser power decreases.

Figure 4.2 shows the final bending angle comparisons between the two cases of different laser power. It is found that when the laser power is decreased to 750 W, the final bending angles along the width direction become smaller than those of 1 000 W. It indicates that the final bending angle decreases with the decrease of the laser power due to the lower peak temperature and the lower temperature gradient.

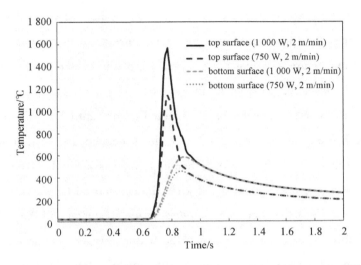

Figure 4.1 Temperature comparisons at the center point of the steel sheet under different laser powers

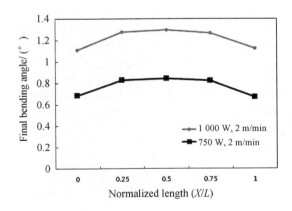

Figure 4.2 Final bending angle comparisons under different laser powers

4.2.2 Effect of Laser Scanning Speed

In order to investigate the effect of the laser scanning speed, a simulation under the conditions in which the laser power is 1 000 W and the scanning speed is 5m/min, is carried out. The temperature comparisons and final bending angle comparisons between this simulation result and the basic simulation (laser power is 1 000 W, scanning speed is 2 m/min) result are shown in Figure 4.3 and Figure 4.4, respectively. It is seen that when the laser scanning speed increases from 2 m/min to 5 m/min while keeping other conditions constant, both the temperature and final bending angle decrease. The reason for this is thought to be the fact that when the laser scanning speed increases, the duration of the heat flux into the material becomes shorter and the temperature peak value and the temperature gradient are both reduced. Another fact is also found from Figure 4.3 that the temperature at the center point of the sheet reaches peak earlier when the laser scanning speed increases.

Furthermore, an interest phenomenon is reflected from Figure 3.8, Figure 4.2 and Figure 4.4 that the final bending angles at the edge of the specimen are smaller than those at the center, for example in Figure 3.8, the final bending angle is about 1.1° when the normalized length is 0 and 1 (edges), but is about 1.3° when the normalized length is 0.5 (center). This is because the heat transfer exists between the edges of the specimen and the ambient air, so that the temperature difference between the top and the bottom surfaces at the edge is smaller than that at the center, which reduced the smaller bending angle at the edge that at the center. This phenomenon is termed edge effect[8, 35, 43-44].

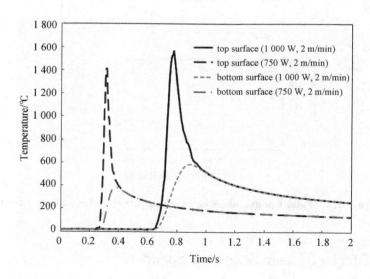

Figure 4.3　Temperature comparisons at the center point of the steel sheet under different scanning speeds

4　Finite Element Analysis of One-time Laser Scanning

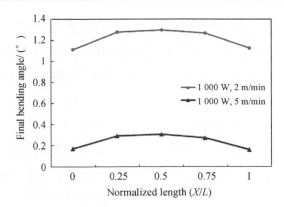

Figure 4.4　Final bending angle comparisons under different scanning speeds

4.2.3　Effect of Line Energy

As mentioned in the previous section, the final bending angle decreases with the decrease of the laser power and with the increase of the laser scanning speed. However, in the experimental work, it is difficult to control the laser power and the scanning speed simultaneously to obtain a required bending angle. Thus, the concept of line energy[2, 36] is proposed, which presents the relationship between laser power and scanning speed. Line energy is defined using the following formulation:

$$B_L = \frac{PA}{v} \tag{4.1}$$

where, E_L—line energy;

　　P—laser power;

　　A—the absorptivity of laser energy;

　　v—laser scanning speed.

The absorptivity A is set as 0.6[37]. Thus, the line energies of the three cases described in the last section can be calculated as follows:

$$B_{L,cace1} = \frac{PA}{v} = \frac{1\ 000 \times 0.6}{2 \times 10^3/60} = 18\ \text{J/mm} \tag{4.2}$$

$$B_{L,cace2} = \frac{PA}{v} = \frac{750 \times 0.6}{2 \times 10^3/60} = 13.5\ \text{J/mm} \tag{4.3}$$

$$B_{L,cace3} = \frac{PA}{v} = \frac{1\ 000 \times 0.6}{5 \times 10^3/60} = 7.2\ \text{J/mm} \tag{4.4}$$

It is found that the line energy of case 1 is the largest (18 J/mm) and that of case 3 is the smallest (7.2 J/mm). The final bending angle comparison among the three cases is shown in Figure 4.5, in which it is seen that the final bending angle in case 1 is the largest (1.27°) and that in case 3 is the smallest (0.24°); this means that the final bending angle decreases with the decrease of the line energy.

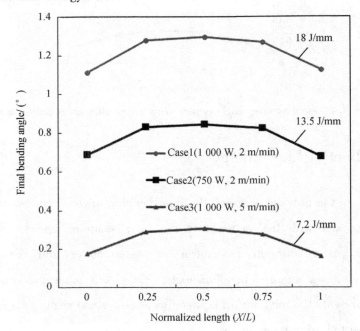

Figure 4.5　Final bending angle comparisons under different line energies

In order to confirm this conclusion, a series of additional simulations with various line energies are performed. The laser power is kept in the range of 750 ~ 1 250 W and the laser scanning speed is 2 ~ 5 m/min. All of the conditions and simulation results are listed in Table 4.1. The relationship between the bending angle and the line energy is shown in Figure 4.6. It is found that the threshold line energy is about 5 J/mm, which means that no deformation occurs if the line energy is below 5 J/mm. Furthermore, when the line energy is above 20 J/mm, the sheet surface melts, meaning that bending deformation is impossible. It indicates that the bending deformation of the steel sheet occurs in the line energy range of 5 ~ 20 J/mm, in which range an almost linear increase of the bending angle is found and the largest bending angle after a single scanning of the laser beam is about 1.27°. According to these results, an appropriate laser power and scanning speed can be selected to obtain the required bending angle in the industrial production.

4 Finite Element Analysis of One-time Laser Scanning

Table 4.1 Final bending angles obtained with various line energies

Laser power/W	Scanning speed/(m·min⁻¹)	Line energy/(J·min⁻¹)	Final bending angle/(°)
750	2	13.5	0.78
	3.5	7.74	0.25
	5	5.4	0
1 000	2	18	1.27
	3.5	10.26	0.53
	5	7.2	0.24
1 250	2	22.5	Mellt
	3.5	12.84	0.72
	5	9	0.41

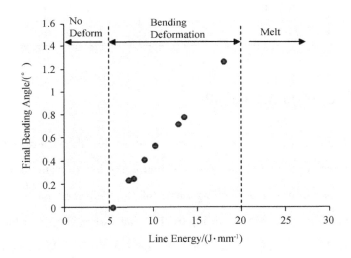

Figure 4.6 Relationship between final bending angles and line energies

4.3 Analysis of Specimen Size Effect

As described previously, the geometry of a specimen also affects the forming behavior of the steel sheet significantly during the laser forming process. In this chapter, the effects of the specimen size, including variation of sheet length, width and thickness, on temperature and final bending angle are investigated. The scanning direction is still along the center line of the metal sheet from point A to point B. The sheet length is defined as the dimension in perpendicular to the laser scanning direction and the sheet width is defined as the dimension in the scanning direction. In the sheet length and width effect analyses, the sheet thickness is fixed at 1.4 mm and the sets of the sheet length and width taken into account are listed in Table 4.2. In the sheet thickness effect analysis, the sheet thickness is changed in the range of 1.0 ~ 2.2 mm. The laser parameters involved are defined as follows: $P = 750$ W, $v = 2$ m/min and $R = 5.2$ mm.

Table 4.2 Sets of sheet length and width for size effect analysis

Length/mm	Width/mm			
	50	100	150	200
50	50 × 50	50 × 100	50 × 150	50 × 200
100	100 × 50			
150	150 × 50			
200	200 × 50			

4.3.1 Effect of Sheet Length

According to Table 4.2, the effect analysis of sheet length is performed by keeping the width of the metal sheet constant (50 mm) and changing the length from 50 mm to 200 mm. The variations of peak temperature and bending angle of the deformed metal sheet are shown in Figure 4.7 and Figure 4.8. It is seen from Figure 4.7 that, when the sheet length increases, the peak temperatures both on the top surface and the bottom surface decrease even though these decreases are very small. The decreases are thinking to be induced by the fact that a larger length provides a larger heat sink[20]. In Figure 4.8, it is seen that the average bending angles increase with the increase of the sheet length. This is because when the sheet length is larger, a larger thermal stress is caused by larger constraint from the cold sheet. This larger thermal stress induces a larger plastic stress which causes a larger plastic deformation. Furthermore, it is found that, when the sheet length is larger than 20 times of the laser beam diameter (5.2 mm), the influences of the sheet length on both the peak temperature and the bending angle are negligible. That's to say, the forming characteristics are getting stable when the sheet length is larger than 20 times of the laser beam diameter.

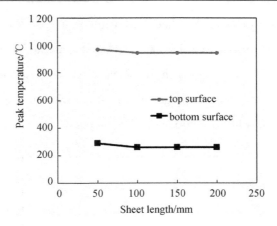

Figure 4.7 Effect of the sheet length on peak temperature

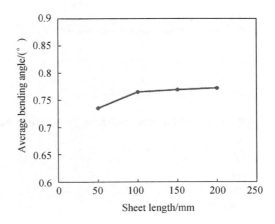

Figure 4.8 Effect of the sheet length on final bending angles

4.3.2 Effect of Sheet Width

Similar to last section, the effect analysis of sheet width is also performed according to Table 4.2, in which the sheet length is kept constant and the sheet width is changed from 50 mm to 200 mm. The variation of peak temperature with the sheet width is shown in Figure 4.9. It is seen that the peak temperatures on both the top surface and the bottom surface drop when the sheet width increases from 50 mm to 100 mm, and then the peak temperatures are getting stable. It is because when the sheet width is small, the laser irradiation ends before the temperature field reaching the so-called quasi-steady state[38] and when the width is larger than 100 mm, the peak temperature does not change after reaching the quasi-steady state.

Figure 4.10 shows the variation of bending angle with the sheet width. It is found that the average bending angle increase with the sheet width increase. As described previously, during the heating process, the thermal expansion of the heated material is restricted by surrounding cool material. This restriction increases with the sheet width which causes a larger thermal stress and this thermal stress is converted to a larger plastic stress when it exceeds the yield strength of the material. Thus, a larger plastic deformation is generated.

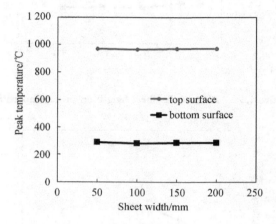

Figure 4.9 Effect of the sheet width on peak temperature

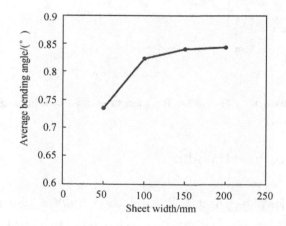

Figure 4.10 Effect of the sheet width on final bending angles

4.3.3 Effect of Sheet Thickness

Figure 4.11 and Figure 4.12 show the variation of peak temperature and bending angle, respectively, with the sheet thickness while the sheet length and width are both kept constant. It is seen from Figure 4.11 that the temperatures on both the top surface and bottom surface decrease with the increase of sheet thickness. Figure 4.12 illustrates that the average bending angle decreases rapidly with the increase of the sheet thickness. That is because when the sheet thickness increases, the thermal expansion of the heated material decreases, which reduced the less plastic stress. Thus, the plastic deformation becomes more difficult and then the final bending angle becomes smaller.

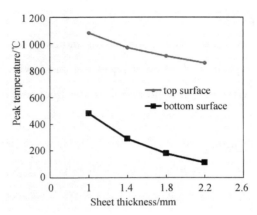

Figure 4.11　Effect of the sheet thickness on peak temperature

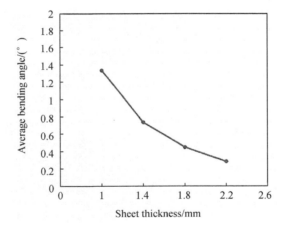

Figure 4.12　Effect of the sheet thickness on final bending angles

4.4　Summary

This chapter has investigated the effects of laser parameters and specimen size on forming characteristics of DP980 high strength steel in the one-time laser scanning process.

In the analysis of laser parameter effect, the results indicate:

- The final bending angle increases with the increase of the laser power, but with the decrease of the laser scanning speed.

- In the line energy range of 5 ~ 20 J/mm, the final bending angle increases linearly with the increase of the line energy. When the line energy is less than 5 J/mm, theenergy is not large enough to bend the steel sheet and when the line energy is larger than 20 J/mm, the material melts due to the high energy.

In the analysis of specimen size effect, the results indicate:

- The final bending angle of DP980 steel sheet increases with the increase of both sheet length and width, but decreases with the increase of sheet thickness, when the laser parameters are kept constant.

- The peak temperature firstly decreases with the increase of both sheet length and width, and then becomes stable as the sheet length and width is getting larger. By contrast, the peak temperature keeps decreasing with the increase of the sheet thickness.

5 Analysis of Scanning Pattern Effect in Multi-time Laser Scanning

5.1 Introduction

The deformation of laser forming process is realized by repeatedly scanning the metal sheet in terms of the small bending angle of each scanning. In the experimental work, the specimen is usually scanned by the laser beam again and again in a single-direction, for example, from point A to point B (see Figure 3.1). However, maybe there is some difference if the scanning direction is changed time by time. Therefore, a series of finite element analyses of different scanning patterns are carried out in this chapter. Firstly, two-time scanning analyses with different scanning patterns are carried out and then three-time scanning analyses are implemented in addition. The specimen dimension employed in the current analysis is 100 mm × 50 mm × 1.4 mm. The laser power is kept on 1 000 W, laser beam diameter is 5.2 mm and the laser scanning speed is 2 m/min.

5.2 Two-time Scanning Analysis

The scanning patterns in the two-time scanning analysis are presented in Table 5.1. It is seen that in case 1, the two times of laser scanning are carried out in the same directions, from point A to point B, as shown in Figure 5.1 (a). But in case 2, the two times of laser scanning are carried out in the opposite directions: the first scanning direction is from point A to point B and the second scanning direction is from B to point A, as shown in Figure 5.1 (b).

Table 5.1 Different scanning patterns in the two-time laser scanning analysis

	First scanning direction	Second scanning direction
Case 1	$A \to B$	$A \to B$
Case 1	$A \to B$	$B \to A$

The final bending angles undergoing the two cases of laser scanning are shown in Figure 5.2. It is found that the average bending angle in case 1 is 1.97° and that in case 2 is 2.048°

which is about 4.0% larger than case 1. Figure 5.3 (a) and (b) show the time histories of temperature for the two cases of analysis. In both of the two cases, the steel sheet is scanned by the laser beam for 2 s and then cooled for 60 s during every scanning process. It is seen that the temperature distributions are almost the same in the two scanning cases. The reason is thought to be that the sheet sizes and the scanning times of the two cases are in full accords.

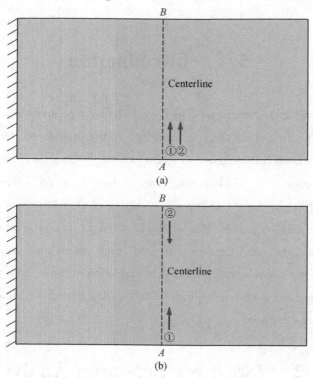

Figure 5.1　Different scanning patterns in the two-time laser scanning analysis

(a) Case 1; (b) Case 2

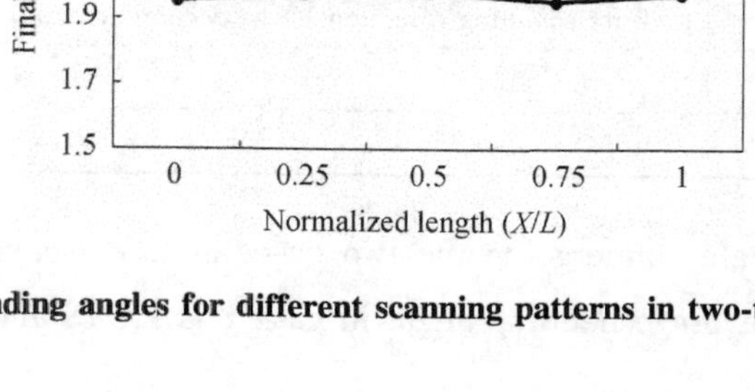

Figure 5.2　Final bending angles for different scanning patterns in two-time scanning analysis

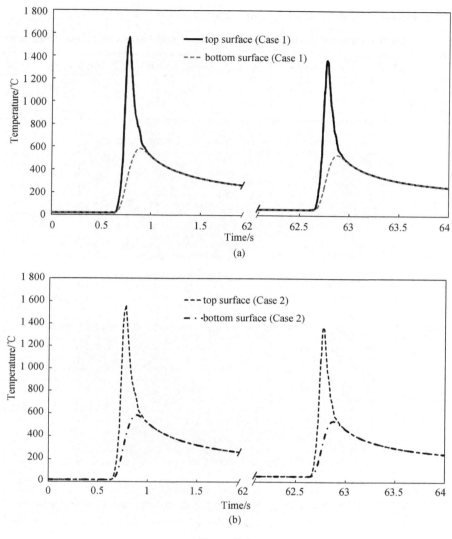

Figure 5.3

(a) Temperature distributions at the center point of the steel sheet in the two-time scanning analysis: Case 1

(b) Temperature distributions at the center point of the steel sheet in the two-time scanning analysis: Case 2

5.3 Three-time Scanning Analysis

In the three-time scanning analysis, two cases of different scanning patterns are also subjected to the analysis, as listed in Table 5.2. One case involved scanning from point A to point B for all of the three times of scanning, as shown in Figure 5.4 (a), and the other case involved scanning from point A to point B for the first and third scanning but from point B to *point A* for the second scanning, as shown in Figure 5.4 (b).

Table 5.2 Different scanning patterns in the three-time laser scanning analysis

	First scanning direction	Second scanning direction	Third scanning direction
Case 1	A→B	A→B	A→B
Case 2	A→B	B→A	A→B

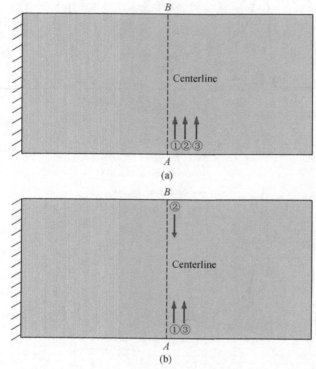

Figure 5.4 Different scanning patterns in the three-time laser scanning analysis
(a) Case 1; (b) Case 2

The final bending angles for the different scanning patterns in the three-time scanning analysis are shown in Figure 5.5. It is seen that the average bending angle in case 2 is 2.684°, which is about 11% larger than that in case 1 (2.414°). The difference of the bending angle in the two-time scanning described in previous section is 4%, which is much smaller than the difference in the three-time scanning (11%). This means that the final bending angle differences among the different scanning patterns become larger with the increase of the scanning times. So, an investigation into the effect of scanning direction is carried out. Figure 5.6 (a) and (b) show the time histories of temperature for the different scanning patterns in the three-time scanning analysis. As described previously, the metal is irradiated by the laser beam for 2 s and cooled for 60 s during every scanning process. It is seen that the temperature distributions are also almost the same for both case 1 and case 2.

5 Analysis of Scanning Pattern Effect in Multi-time Laser Scanning

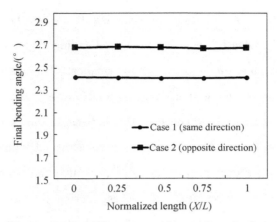

Figure 5.5 Final bending angles for different scanning patterns in three-time scanning analysis

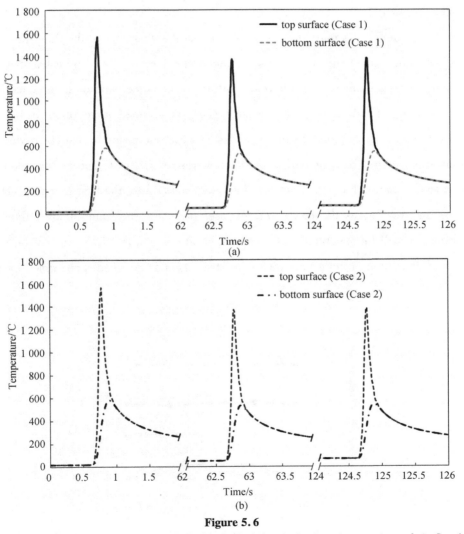

Figure 5.6
(a) Temperature distributions at the center point of the steel sheet in the three-time scanning analysis: Case 1
(b) Temperature distributions at the center point of the steel sheet in the three-time scanning analysis: Case 2

In addition, the effect of the different scanning times on the final bending angle is investigated in Figure 5.7. It can be seen that the final bending angle of the three-time scanning is the largest and that of the one-time scanning is the smallest. This means that the final bending angle also increases as the scanning times increases. However, the bending angle does not increase linearly as the scanning time increases. This is because the existing residual stress after the first scanning results in the smaller stresses in the second and following scanning processes when the parameters of laser beam keep constant[39]. So that the bending angle increment is getting smaller and smaller time by time. Moreover, during the second scanning, the heat affected zone (HAZ) is getting much larger than that in the first scanning, which may affect the bending behavior also[39-40].

On the other hand, it is found from Figure 5.7 that the curvature of the three-time scanning is the smallest, but that of the one-time scanning is the largest. This indicates that the final bending angle difference among the different positions of the sheet surface is the smallest in the three-time scanning. That's to say, if the steel sheet is scanned more times using the laser beam, the difference of the final bending angles at different positions can be reduced. This phenomenon owes to the residual stress of the material which is larger at the center (large deformation zone) than at the edge (small deformation zone) after the first scanning. During the second scanning process, as the parameters of laser beam are the same all over the length, a larger deformation will be generated at the edge than at the center. Thus, it can be concluded that bending deformation is more uniform if the steel sheet is scanned more times by the laser beam.

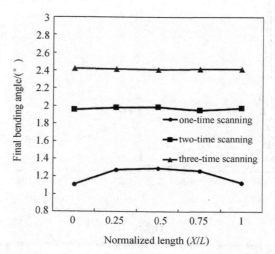

Figure 5.7 Final bending angle comparisons for different scanning times

5.4 Summary

In this chapter, the multi-scanning analysis of DP980 high strength steel under the same laser parameters and the same specimen size has been carried out. The result shows that the final bending angle increases with the increase of the scanning times, but it is a nonlinear increase due to the existing residual stress after the first time scanning. Furthermore, the bending deformation is getting more and more uniform with the increase of the scanning times.

The most important, it has been found that in the multi-scanning process, scanning the sheet metal with diverse path direction at the adjacent two scanning times obtains a larger bending angle than that using the same path direction every time.

6 Conclusions and Recommendations

6.1 Conclusions

This paper has investigated the laser forming characteristics of the DP980 high strength steel using Gaussian heat source model via the finite element analysis. The main achievements of the work described in this paper are summarized as follows:

• A Gaussian laser source model is established for the finite element analysis of laser forming process using ABAQUS user subroutine DFLUX. The finite element analysis result for the DP980 high strength steel sheet is confirmed valid through the verification of experiment.

• The effects of laser parameters on forming characteristics are investigated that the final bending angle increases with the increase of the laser power, but with the decrease of the laser scanning speed. In the line energy range of 5 ~ 20 J/mm, the final bending angle increases linearly with the increase of the line energy.

• The effects of specimen size on forming characteristics are also illustrated as that the final bending angle of DP980 steel sheet increases with the increase of both sheet length and width, but decreases with the increase of sheet thickness, when the laser parameters are kept constant.

• In the multi-scanning process of DP980 high strength steel, the final bending angle increases nonlinearly with the increase of the scanning times and the bending deformation is getting more and more uniform all over the specimen. Furthermore, scanning the sheet metal with diverse path direction at the adjacent two scanning times can obtain a larger bending angle than that using the same path direction every time.

6.2 Recommendations

Through the research work on laser forming of high strength steel, it is considering that a few of potential works for improving the research level of the laser forming technique are recommendable.

6.2.1 Finite Element Models

In the actual experiments, when the laser energy is very large, the material undergoing the laser irradiation melts and the phase changes. However, the melting process and the phase changing phenomena are not accounted in the finite element analysis of the current study. Completing the finite element models considering the material changes in the simulations could result in a gain in accuracy.

6.2.2 Simulation of Complex Shapes

This paper has only presented the simulations of a straight line irradiation for a rectangular specimen by the laser beam. It is worthy to do further simulations of some complex scanning paths of the laser beam and more complex shapes of specimen.

Appendix A Overview of Laser

A.1 Properties of Laser

Laser is the acronym of Light Amplification by Stimulated Emission of Radiation. The key words in the acronym are "amplification by stimulated emission". A laser amplifies light and it does this by means of the phenomenon of stimulated emission. The theory of laser derived early from the famous physicist Albert Einstein in 1917. When the particles on the high energy level of a matter encounters a photon, they jump to the low energy level while radiating a light with the same properties with the photon. The light moves through space as a wave. The energy of photon is $E = hf$ where h is Planck's constant and f is the wavelength of the light.

Due to the release of any photons by an atom, laser has the unique properties of a wavelength, phase and energy level dependent on the difference between the excited atom and the ground-state energy level[2]. This properties make lasers such useful tools in an incredible variety of industrial applications. The primary properties associated with manufacturing applications are radiance, monochromaticity, coherence and the various forms of output.

Radiance of a source of light is the power emitted per unit area of the source per unit solid angle. Lasers achieve high radiance at relatively low power levels. The amount of power that can be concentrated on a spot by focusing a beam of light is directly proportional to the radiance. In laser work, the irradiance (frequently called intensity) is the power per unit area falling on a surface at a given point. For convenience, the units are usually given as watts per square centimeter (W/cm^2).

Monochromatic means single color or single wavelength. Actually, no light source is perfectly monochromatic. Lasers tend to be relatively monochromatic. One specific wavelength of light is determined by the amount of energy released when the electron drops to a lower-energyorbital.

Lasers provide a high radiance source of light with a high degree of coherence. Coherence means each photon moves in step with the other (i. e., all photons have wave fronts that move in unison).

The output of industrial lasers may be pulsed or continuous. Materials processing lasers may be either pulsed or continuous whereas lasers used for alignment, measurement and other low power applications usually have a continuous output. Continuous output is referred to as continuous wave (CW) output. However, it usually simply means that the power output is constant. Pulsed output may be accomplished by a variety of techniques. By using different pulse generation techniques the pulse duration, pulse energy and reproducibility can be modified over wide ranges.

A.2 Types of Laser

There are many different types of lasers. Laser medium can be a solid, gas, liquid or semiconductor. Lasers are commonly designated by the type of lasing material employed.

Solid-state lasers have lasing material distributed in a solid matrix (such as the ruby or neodymium: yttrium-aluminum garnet "Yag" lasers). The neodymium—Yag laser emits infrared light at 1 064 nm.

Gas lasers (helium and helium-neon, HeNe, are the most common gas lasers) have a primary output of visible red light. CO_2 lasers emit energy in the far-infrared, and are used for cutting hard materials.

Excimer lasers, whose name is derived from the terms excited and dimmers, use reactive gases, such as chlorine and fluorine, mixed with inert gases such as argon, krypton or xenon. When electrically stimulated, a pseudo molecule (dimmer) is produced. When lased, the dimmer produces light in the ultraviolet range.

Dye lasers use complex organic dyes, such asrhodamine 6G, in liquid solution or suspension as lasing media. They are tunable over a broad range of wavelengths.

Semiconductor lasers, sometimes called diode lasers, are not solid-state lasers. Theses electronic devices arefenerally very small and use low power. They may be built into larger arrays, such as the writing source in some laser printers or CD players.

Appendix B Lists of ABAQUS Computer Program

The following pages contain the lists of ABAQUS computer programs for the finite element simulation referred in chapter 3 and user subroutines for the Gaussian-distributed heat source employed in the 3-time laser scanning analysis.

B.1 Finite Element Analysis of One-time Laser Scanning Process

```
* Heading
** Job name: 1000w-2mpm-100+50 Model name: 1000w-A08-980-top surface-2mpm-100+50
** Generated by: Abaqus/CAE 6.10-1
* Preprint, echo=NO, model=NO, history=NO, contact=NO
**
** PARTS
**
* Part, name=Part-1
* End Part
**
**
** ASSEMBLY
**
* Assembly, name=Assembly
**
* Instance, name=Part-1-1, part=Part-1
* Node
1, 0.00999999978, 0.0250000004, 0.
```

2, 0.00999999978, -0.0250000004, 0.
3, 0.0500000007, -0.0250000004, 0.
4, 0.0500000007, 0.0250000004, 0.
5, -0.00999999978, 0.0250000004, 0.
6, -0.00999999978, -0.0250000004, 0.
7, -0.0500000007, 0.0250000004, 0.
8, -0.0500000007, -0.0250000004, 0.
9, 0.00999999978, 0.0240000002, 0.
10, 0.00999999978, 0.023, 0.

1470, -0.0399999991, 0.0149999997, 0.
1471, -0.0399999991, 0.0160000008, 0.
1472, -0.0399999991, 0.0170000009, 0.
1473, -0.0399999991, 0.0179999992, 0.
1474, -0.0399999991, 0.0189999994, 0.
1475, -0.0399999991, 0.0199999996, 0.
1476, -0.0399999991, 0.0209999997, 0.
1477, -0.0399999991, 0.0219999999, 0.
1478, -0.0399999991, 0.023, 0.
1479, -0.0399999991, 0.0240000002, 0.
* Element, type = S4RT
1, 1, 9, 255, 112
2, 9, 10, 256, 255
3, 10, 11, 257, 256
4, 11, 12, 258, 257
5, 12, 13, 259, 258
6, 13, 14, 260, 259
7, 14, 15, 261, 260
8, 15, 16, 262, 261
9, 16, 17, 263, 262
10, 17, 18, 264, 263

1390, 1469, 1470, 212, 213
1391, 1470, 1471, 211, 212
1392, 1471, 1472, 210, 211
1393, 1472, 1473, 209, 210
1394, 1473, 1474, 208, 209
1395, 1474, 1475, 207, 208
1396, 1475, 1476, 206, 207
1397, 1476, 1477, 205, 206
1398, 1477, 1478, 204, 205
1399, 1478, 1479, 203, 204
1400, 1479, 202, 7, 203
* Nset, nset = _PickedSet3, internal, generate
1, 1479, 1
* Elset, elset = _PickedSet3, internal, generate
1, 1400, 1
** Section: Section - 1
* Shell Section, elset = _PickedSet3, material = STEEL, offset = SPOS
0.0014, 5
* End Instance
**
* Nset, nset = __PickedSet15, internal, instance = Part-1-1
7, 8, 203, 204, 205, 206, 207, 208, 209, 210, 211, 212, 213, 214, 215, 216
217, 218, 219, 220, 221, 222, 223, 224, 225, 226, 227, 228, 229, 230, 231, 232
233, 234, 235, 236, 237, 238, 239, 240, 241, 242, 243, 244, 245, 246, 247, 248
249, 250, 251
* Elset, elset = __PickedSet15, internal, instance = Part-1-1, generate
1351, 1400, 1
* Nset, nset = __PickedSet44, internal, instance = Part-1-1, generate
1, 1479, 1
* Elset, elset = __PickedSet44, internal, instance = Part-1-1, generate
1, 1400, 1
* Elset, elset = __PickedSurf41__SPOS, internal, instance = Part-1-1, generate

1, 1400, 1
* Surface, type = ELEMENT, name = _ PickedSurf41, internal
_ PickedSurf41 _ SPOS, SPOS
* Elset, elset = _ PickedSurf42 _ SPOS, internal, instance = Part-1-1, generate
1, 1400, 1
* Surface, type = ELEMENT, name = _ PickedSurf42, internal
_ PickedSurf42 _ SPOS, SPOS
* Elset, elset = _ PickedSurf43 _ SPOS, internal, instance = Part-1-1, generate
1, 1400, 1
* Surface, type = ELEMENT, name = _ PickedSurf43, internal
_ PickedSurf43 _ SPOS, SPOS
* End Assembly
**
** MATERIALS
**
* Material, name = STEEL
* Conductivity
50.32, 22.
50.32, 76.0563
51.12, 143.662
48.72, 202.817
46.33, 261.972
43.93, 352.113
41.53, 402.817
38.34, 476.056
35.14, 552.113
30.35, 650.704
26.36, 777.465
25.56, 850.704
24.76, 923.944
24.76, 1000.
24.76, 1050.7

24.76, 1109.86
24.76, 1152.11
26.36, 1222.54
28.75, 1295.77
29.55, 1356.06
33.55, 1445.07
35.94, 1501.41
35.14, 1597.18
35.14, 1701.41
35.14, 1800.
* Density
7800.,
* Elastic
2.1e+11, 0.29, 20.
1.6e+11, 0.32, 500.
1e+11, 0.35, 1000.
* Expansion
1.52e-05,
* Plastic
5.4e+08, 0., 20.
7.7e+08, 0.05, 20.
8.5e+08, 0.1, 20.
8.9e+08, 0.15, 20.
9.2e+08, 0.2, 20.
4.2e+08, 0., 200.
5.6e+08, 0.05, 200.
6.8e+08, 0.1, 200.
7e+08, 0.15, 200.
7.5e+08, 0.2, 200.
3.5e+08, 0., 400.
5e+08, 0.05, 400.
5.6e+08, 0.1, 400.

5.8e+08, 0.15, 400.

6e+08, 0.2, 400.

*Specific Heat

469.142, 22.

469.142, 77.7778

485.319, 150.

505.541, 219.444

509.585, 275.

537.895, 350.

570.25, 400.

602.605, 458.333

630.915, 519.444

667.314, 577.778

687.535, 616.667

711.801, 647.222

849.308, 658.333

962.55, 672.222

1091.97, 675.

1253.74, 680.556

1415.51, 686.111

1569.2, 691.667

1726.93, 694.444

1856.35, 702.778

1767.37, 730.556

1597.51, 733.333

1455.96, 736.111

1253.74, 738.889

1075.79, 741.667

946.372, 744.444

792.688, 750.

679.447, 761.111

570.25, 772.222

578.339, 827.778
590.472, 897.222
598.56, 952.778
614.738, 1027.78
622.826, 1125.
639.003, 1200.
655.181, 1247.22
651.136, 1305.56
663.269, 1366.67
934.239, 1369.44
1168.81, 1377.78
1419.56, 1383.33
1654.13, 1388.89
1836.12, 1391.67
2026.21, 1397.22
2034.3, 1477.78
2050.47, 1530.56
2058.56, 1600.
703.713, 1616.67
711.801, 1677.78
719.89, 1741.67
727.979, 1800.
**
** PHYSICAL CONSTANTS
**
* Physical Constants, absolute zero = -273.15, stefan boltzmann = 5.6704e-08
**
** BOUNDARY CONDITIONS
**
** Name: BC-1 Type: Symmetry/Antisymmetry/Encastre
* Boundary
_ PickedSet15, ENCASTRE

**
** PREDEFINED FIELDS
**
** Name: Predefined Field-1 Type: Temperature
* Initial Conditions, type = TEMPERATURE
_ PickedSet44,20. , 20. , 20. , 20. , 20.
** --
**
** STEP: Step-1
**
* Step, name = Step-1, nlgeom = YES, inc = 10000
* Coupled Temperature-displacement, creep = none, deltmx = 1000.
0.05, 2. , 2e−05, 2.
**
** LOADS
**
** Name: Load-1 Type: Surface heat flux
* Dsflux
_ PickedSurf43, SNU, 1.
**
** INTERACTIONS
**
** Interaction: Int-1
* Sfilm
_ PickedSurf41, F,20. , 10.
** Interaction: Int-2
* Sradiate
_ PickedSurf42, R,20. , 0.85
**
** OUTPUT REQUESTS
**
* Restart, write, frequency = 1

**
** FIELD OUTPUT: F-Output-1
**
* Output, field, variable = PRESELECT
**
** HISTORY OUTPUT: H-Output-1
**
* Output, history, variable = PRESELECT
* End Step
** ---
**
** STEP: Step-2
**
* Step, name = Step-2, nlgeom = YES, inc = 10 000
* Coupled Temperature-displacement, creep = none, deltmx = 1 000.
6. ,60. , 3e −05, 60.
**
** LOADS
**
** Name: Load-1 Type: Surface heat flux
* Dsflux, op = NEW
**
** OUTPUT REQUESTS
**
* Restart, write, frequency = 1
**
** FIELD OUTPUT: F-Output-1
**
* Output, field, variable = PRESELECT
**
** HISTORY OUTPUT: H-Output-1
**

* Output, history, variable = PRESELECT

* End Step

B.2 User Subroutine for the Gaussian-distributed Heat Source Model

```
SUBROUTINEDFLUX( FLUX, SOL, JSTEP, JINC, TIME, NOEL, NPT, COORDS, JLTYP, 1
TEMP,PRESS,SNAME)
C
INCLUDE 'ABA _ PARAM. INC'


DIMENSIONCOORDS(3),flux(2),time(2)
CHARACTER * 80 SNAME

    pi = 3.1416
    p = 1000
    v = 0.0333
    d1 = v * time(1)
    eff1 = 0.6
    eff2 = 0.5
    eff3 = 0.5
    x = COORDS(1)
    y = COORDS(2)
    z = COORDS(3)

    rad = 0.0026
C rad is the radius of heat source


if( JSTEP. eq. 1 ) then
x0 = 0
```

```
        y0 = -0.025
        z0 = 0
heat = (2 * eff1 * p)/(pi * rad ** 2)
r = (x - x0) ** 2 + (y - y0 - d1) ** 2
shape = exp(-2 * r/rad ** 2)
JLTYP = 0
flux(1) = heat * shape
endif
if(JSTEP. eq. 3) then
x0 = 0
        y0 = 0.025
        z0 = 0
heat = (2 * eff2 * p)/(pi * rad ** 2)
r = (x - x0) ** 2 + (y0 - y - d1) ** 2
shape = exp(-2 * r/rad ** 2)
JLTYP = 0
flux(1) = heat * shape
endif
if(JSTEP. eq. 5) then
x0 = 0
        y0 = -0.025
        z0 = 0
heat = (2 * eff3 * p)/(pi * rad ** 2)
r = (x - x0) ** 2 + (y - y0 - d1) ** 2
shape = exp(-2 * r/rad ** 2)
JLTYP = 0
flux(1) = heat * shape
endif
RETURN
END
```

References

[1] KITAMURA N. Technical report of joint project on materials processing by high power laser. JWES-TP-8302, 1983: 359-371.

[2] JUNG H C. A study on laser forming processes with finite element analysis. Christchurch: University of Canterbury, 2006.

[3] GEIGER M, MEYER-PITTROFF F. Laser beam bending of metallic foils. Proceedings of SPIE, 2002, 4426: 187-190.

[4] VOLLERTSEN F. Forming and Rapid Prototyping. Handbook of the Euro Laser Academy, Schuocker, D. Eitor, Chapman & Hall: Vienna, Austria, 1998.

[5] QAISER N. A study on designing of 3D-laser forming process for circular sheet metal. KAIST, South Korea, 2010.

[6] GEIGER M, ARNET H, VOLLERTSEN F. Laser forming. Proceedings of the LANE, 1994: 81-92.

[7] THOMSON G, PRIDHAM M S. Controlled laser forming for rapid prototyping. Rapid Prototyping Journal, 1997, 3(4): 137-143.

[8] MAGEE J, WATKINS K G, STEEN W M. Laser forming of aerospace alloys. Proceedings of the 1997 Laser Materials Processing Conference, 1997: 156-165.

[9] EDWARDSON S P, et al. 3D laser forming of saddle shapes, Laser Assisted Net Shape Engineering. Proceedings of the LANE, 2001: 559-568.

[10] HENNIGE T. Laser forming of spatially curved parts. Proceedings of the LANE, 1997: 409-420.

[11] OJEDAA C V, GREZ J R. Bending of stainless steel thin sheets by a raster scanned low power CO_2 laser. Journal of Materials Processing Technology, 2009, 209: 2641-2647.

[12] VOLLERTSEN F. Mechanisms and models for laser forming, Laser Assisted Net Shape Engineering. Proceedings of the LANE, 1994: 345-359.

[13] VOLLERTSEN F, RODLE M. Model for the temperature gradient mechanism of laser bending. Proceedings of the LANE, 1994: 371-378.

[14] ARNET H, VOLLERTSEN F. Extending laser bending for the generation of convex

shapes. Proceeding of the Institution of Mechanical Engineering, 1995, 209: 433 -442.
[15] VOLLERTSEN F. An analytical model for laser bending. Lasers in Engineering, 1994, 2: 261 - 276.
[16] CHENG P J, LIN S C. An analytical model for the temperature field in the laser forming of sheet metal. Journal of Materials Processing Technology, 2000, 101: 260 - 267.
[17] CHENG P J, LIN S C. An analytical model to estimate angle formed by laser. Journal of Materials Processing Technology, 2001, 108: 314 - 319.
[18] JI Z, WU S. FEM simulation of temperature field during the laser forming of sheet metal. Journal of Materials Processing Technology, 1998, 74: 89 - 95.
[19] LI L Q, CHEN Y B. FEM simulation for laser forming processing. Acta Metallurgica Sinica (English Letters), 2004(713): 317 - 322.
[20] CHENG P, et al, Analysis and prediction of size effect on laser forming of sheet metal. Transactions of NAMRI/SME, 2004, 32: 28 - 41.
[21] ZHANG J. Effects of scanning schemes on laser tube bending. Transactions of the ASME, 2006, 128: 20 - 33.
[22] GUAN Y J, et al. Finite element modeling of laser bending of pre-loaded sheet metals. Journal of Materials Processing Technology, 2003, 142: 400 - 407.
[23] SONG J H, et al. Bending Characteristics of DP980 Steel Sheets by the Laser Irradiation. Transactions of Materials Processing, 2012, 21(4): 378 - 383.
[24] EDWARDSON S P, et al., Laser forming of fiber metal laminates. Lasers in Engineering, 2005, 15: 233 - 255.
[25] CAREY C, et al, Low power laser forming of glass fiber based fiber metal laminates. Laser Assisted Net Shape Engineering, Proceedings of the LANE, 2007, 5: 645 - 655.
[26] SHEN H, SHI Y, YAO Z. Numerical simulation of the laser forming of plates using two simultaneous scans. Computational Materials Science, 2006, 37: 239 - 245.
[27] VOLLERTSEN F, KOMEL I, KALS R. The laser bending of steel foils for microparts by the buckling mechanism-a model. Modelling and Simulation in Materials Science and Engineering, 1995, 3(1): 107 - 119.
[28] KRAUS J. Basic processes in laser bending of extrusions using the upsetting mechanism, Laser Assisted Net Shape Engineering 2. Proceedings of the LANE, 1997, 2: 431 - 438.

[29]　陈敦军,吴诗惇,向毅斌,等. 板料激光成形的机制及其应用. 兵器材料科学与工程,2000, 23(6): 58 - 61.

[30]　管延锦,孙胜. 板料激光弯曲的屈曲机理的研究. 激光技术,2001,25(1): 11 - 14.

[31]　王秀凤,JANOS T, GYORGY K. 激光弯曲机理的实验研究. 北京航空航天大学学报,2002,28(4): 473 - 476.

[32]　尚晓峰,苏荣华,王志坚. 激光弯曲成形及功能梯度材料成形技术. 北京:冶金工业出版社,2013.

[33]　ABAQUS software, 2007, Version 6.10, SIMULIA Inc., France.

[34]　VENKADESHWARAN K, DAS S, MISRA D. Finite element simulation of 3-D laser forming by discrete section circle line heating. International Journal of Engineering, Science and Technology, 2010, 2(4): 163 - 175.

[35]　BAO J, YAO Y L. Analysis and prediction of edge effects in laser bending. Journal of Manufacturing Science and Engineering, 2001, 123(1): 53 - 61.

[36]　YAU C L, Chan K C, LEE W B. Laser bending of leadframe materials. Journal of Materials Processing Technology, 1998, 82 (1/2/3): 117 - 121.

[37]　KO D C, LEE C J, KIM B M. Production of CO_2 laser forming machine for bending of sheet metal using the FE - analysis. Transactions of Materials Processing: Journal of the Korean Society for Technology of Plastics, 2006 (514): 318 - 325.

[38]　ZHANG L W, et al. FEM simulation of laser forming process of shipbuilding steel plate. Journal de Physique Archives, 2004, 120: 507 - 512.

[39]　FENG S X. Laser bending and forming technology of functionally graded materials: Chinese Edition. Metallurgical Industry, 2012: 108 - 113, 136 - 140.

[40]　WANG H Y, LI C, LUO G. Controlling the heat affected zone (HAZ) in HF pipe and tube welding. Welded Pipe and Tube, 2012, 35(4): 66 - 70.

[41]　KERMANIDIS TH B, KYRSANIDI AN K. Numerical simulation of the laser forming process in metallic sheet metals. Proceedings of the International Conference on Computer Methods and Experimental Measurements for Surface Treatment Effects, 1997: 307 - 316.

[42]　KYRSANIDI AN K. Numerical and experimental investigation of the laser forming process. Journal of Materials Processing Technology, 1999, 87: 281 - 290.

[43]　MAGEE J, WATKINS K G, STEEN W M. Advances in laser forming. Journal of Laser Applications, 1998, 10(6): 235 - 246.

[44] BAO J C, YAO Y L. Study of edge effects in laser bending. American Society of Mechanical Engineers, 1999: 941-948.

[45] HUEBNER K H, et al. The finite element method for engineers. 4th ed. New York: John Wiley & Sons, 2001.

[46] MEINDERS T. Developments in numerical simulations of the real-life deep drawing process. [S.l.]: University of Twente, 1999.

[47] MARCINIAK Z, DUNCAN J L, HU S J. Mechanics of sheet metal forming, 2002.

[48] 尚福林. 塑性力学基础. 西安:西安交通大学出版社, 2011.

[49] WAGONER R H, CHENOT J L. Fundamentals of metal forming. New York: John Wiley & Sons, 1997.

[50] INCROPERA F P, DEWITT D P. Fundamentals of heat and mass transfer. 5th ed. New York: John Wiley & Sons, 2002.

[51] SELVADURAI A P S. Partical differential equations in mechanics 2. Belin: Springer-Verlag, 2000.

[52] CHKRABARTY J. Theory of plasticity. 2nd ed. [S.l.]: McGraw-Hill, 1998.

[53] HUANG Y, LEU D. Finite element analysis of contact problems for a sheet metal bending process. Journal of Computers and Structures, 1995, 57(1): 15-27.

Acknowledgement

On the completion of this paper, I would like to express my deepest gratitude to all who have helped me make the paper possible and better.

Firstly, I am deeply grateful to my honorable supervisor, Professor Jong Bong Kim, who has checked through my papers with patience and given me valuable suggestions, and also improved me in language. His effective advice and comment indicated a bright road in my future writing. Then, I would like to thank theDr. s in KITECH. Dr. Nak Kyu Lee, who has offered me a great opportunity to obtain the research experience and study more in KITECH. Dr. Hye Jin Lee, who is always kind for me and has helped me so much not only in research but also in life since I came to KITECH in 2009. Dr. Jean Ho Park has given me so many supports in work and also shared the parenting experiences with me. Dr. Jung Han Song, who has taught me so much in study and brought me to the laser forming research and finally complete my paper. I would also like to thank the administrative staffs of KITECH, Mrs. Kea Yoon Lee, Soo Mi Lee, Hee Su An and Enu Su Junk, for their actively assist on extending my visa in Korea as a foreign student. Then I would like to express my gratitude to my sincere friend, Aruem Han, who has taken care of me with her kindness not only in work, but also in life, like an elder sister even though we are of the same age. In addition, thank my team members, Si Mo Yeon and Seung Woo Chun for taking care of me during I was working in KITECH.

Finally, I shall to extend my thanks to my family. My husband Wenbin Dong has been the greatest support of me during I was working for my paper. He cooked delicious food for me every day and massaged for me when I was tired and looked after me in all ways. Besides, he looked after my daughter by himself when I was working. Without his dedication, I could not complete my paper. Then, my father Xianyou Zhang and mother Fengzhi Li always ask me to concentrate on my study and not to worry about them. And they have always encouraged me and given me confidence when I was in trouble. Also, I would like to thank my brother Jianzheng Zhang, sister - in - law Guimei Li and my little nephew Zihe. They have been looking after our parents well and made me reassuring during I am studying in Korea. Last but not least, I would like to express my gratitude to my father - in - law Hua Dong and mother - in - law Yaping Li. They have taken care of me after my daughter was born and then they have been looking after my little daughter for me for more than two years in order to make me continue my paper. And my lovely daughter Zhiqiao, who is also the one I would like to thank, has always been the power to make me improve and finally complete the paper.